From Unstable to Unstoppable: The Rise of Organic Free Radicals in Chemistry

Meenu

Table of Contents

Chapter 1

Introduction to Radical-Based Molecular Materials

1.1 Opening Remarks

Organic free radicals, comprised of first-row p-block elements, are defined as molecular species with subvalent electronic structures. For these systems, the open valence introduces an intrinsic instability that warrants a degree of stabilization.[1] Most often, this is translated to the reactivity of radicals, in which a thermodynamic gain is achieved through the spontaneous pairing of electrons, either by atom or electron transfer processes (e.g., dimerization, disproportionation reaction, redox reaction), requiring little to no activation energy.[2] Under these circumstances, molecular radicals are short-lived and transient in nature. In other instances, molecular design strategies can be used to preserve the open-shell character, affording organic radicals with longer lifetimes.[3–5] Generally, these approaches invoke a kinetic or thermodynamic stability to the electronic structure. Regarding the former, methods utilize steric protection to inhibit the unpaired electron from reacting with its environment, while the latter involves the distribution of spin density over the molecular framework thereby lowering the potential for reactivity. In many examples, both approaches are employed to modulate the chemical stability of radicals, depending on the application at hand, and with that, radicals have received popularity towards roles in molecular transformations,[2,6] biological applications[7,8] and functional materials.[3,9]

Interestingly, the consideration that radicals could be stable was unconceivable just over a century ago. In fact, chemists struggled to accept the existence of organic radicals, as characterization of these materials was a near impossible feat.[10] This notion, however, had been reconsidered upon the discovery of the triphenylmethyl radical **1-1** by Moses Gomberg in 1900.[11] His initial attempt to synthesize hexaphenylethane through a reaction involving triphenylmethyl bromide with silver metal resulted in the immediate precipitation of the peroxide **1-2** (Scheme 1.1). Gomberg realized the exclusion of oxygen from this reaction halted precipitation until it was reintroduced into the system. Experimental data exclusively collected through analytical and calorimetric methods supported the elucidation of **1-1** as a radical. This report undoubtably sparked the renaissance of radical chemistry within the 20th century, which not only supported the existence of free radicals but shed new light on the stability of open-shell molecules.

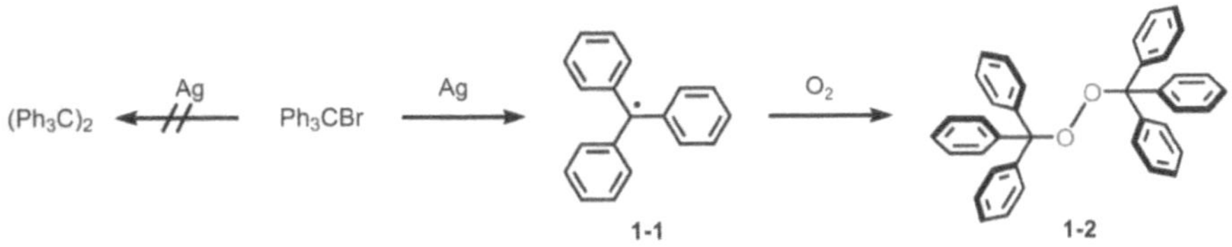

$(Ph_3C)_2 \xleftarrow{\text{Ag}}\!\!\!/\!\!\!/\!\!\!\!\!\!- Ph_3CBr \xrightarrow{\text{Ag}}$ **1-1** $\xrightarrow{O_2}$ **1-2**

Scheme 1.1 Synthesis and oxidation of triphenylmethyl radical **1-1**.

1.2 Modulation of Radical Stability Through Molecular Design Strategies

1.2.1 Stability Enhancement *via* Steric Protection

With advancements in technology, the stability and electronic structure of radicals have been studied extensively. It should be stated that the term "stability" is subjective in respect to radicals and bears no universal definition. However, for the purpose of this book, the definition inscribed by Griller and Ingold will be used to bring context to radical stability.[12] A radical deemed "persistent" infers that its lifetime is long enough to be detected by spectroscopic methods but cannot be isolated in a pure state, whereas the descriptive term "stable" implies a radical that is unreactive to air and moisture in solution and solid state indefinitely. In that regard, triphenylmethyl (**1-1**) represents the first account of a persistent radical.[11] Accordingly, **1-1** has not been isolated discretely in the solid state, but rather self-associates forming the σ-bonded dimer **1-3** (Chart 1.1).[13] Nonetheless, this C—C σ-bond is relatively weak (dissociation enthalpy = 45 kJ·mol^{-1}),[14] such that dissolution of **1-3** re-establishes an equilibrium with the monomer **1-1**. The reversibility of dimerization, and thus persistent nature of the neutral radical in a deoxygenated solution can be attributed to the steric congestion brought about by the phenyl groups.[15]

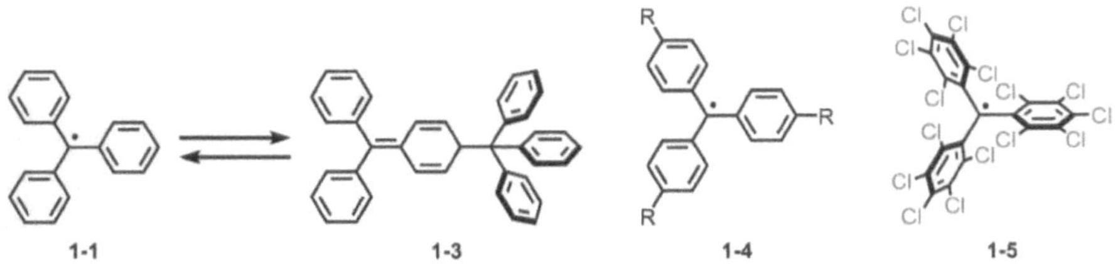

1-1   **1-3**   **1-4**   **1-5**

Chart 1.1 Self-association of triphenylmethyl radical **1-1** along with examples of sterically protected derivatives.

From a kinetic standpoint, the open-shell structure of **1-1** remains susceptible towards reactivity with its surroundings due to a degree of spin delocalization along the phenyl π-system, as demonstrated by the propensity of **1-1** to dimerize (**1-3**) or oxidize (**1-2**). This vulnerability can be circumvented through additional shielding measures *via* chemical modifications along the phenyl moieties (e.g., **1-4** and **1-5** in Chart 1.1). For instance, σ-dimerization can be alleviated through functionalization at the *para*-position of the phenyl rings blocking reactivity at these peripheral sites (e.g., **1-4**).[16,17] Moreover, when *ortho*-substituents are also introduced, rotational mobility about the aryl moieties is reduced from the repulsive van der Waals forces generated by these groups and consequentially, spin density is concentrated on the central sp^2-hybridized carbon atom. This effect becomes more pronounced upon chlorination of the phenyl groups, as in **1-5**, in which the unpaired spin is confined to this center.[18,19] Essentially, this is due to the perchlorophenyl rings locked into one of two propeller-shaped conformations as a result of a high intrarotational barrier (96 kcal mol⁻¹).[20–22] In so doing, the voids in-between the protruding aryl moieties are filled by the size of the *ortho*-chlorine atoms, which further shields the spin density of **1-5** against most chemical reactivity, both in solution and the solid state.[19]

1.2.2 Stability Enhancement *via* Delocalization

In sharp contrast to steric protection, thermodynamic approaches aim to stabilize open-shell structures through electronic effects intrinsic to the molecule. One of the most common tactics is the delocalization of the unpaired electron through a conjugated π-framework. In this scenario, spin density is divided among a manifold of atoms and adequately reduces reactivity at a given location. A prime example demonstrating this effect is the phenalenyl radical **1-6**; a fused carbon-based polycyclic framework devoid of steric protection.[23–25] The rigid platform enables distribution of electron density in the singly occupied molecular orbital (SOMO) to the six alpha-carbon positions equivalently. In other words, the spin density of the single electron is evenly divided between the carbon atoms. The effect of resonance stabilization is apparent in solution, in which **1-6** can exist for an extended period; however, it should be noted that such stability only occurs under strict anaerobic conditions at room temperature. At lower temperatures, **1-6** succumbs to its kinetic instability where an equilibrium forms between the radical and its σ-dimer **1-7** in solution.[26] In the solid state, the radical dimerizes entirely and, under aerobic conditions, rapidly decomposes to peropyrene (**1-8**; Scheme 1.2).[23,27]

(a)

(b)

1-6   1-7   1-8

Scheme 1.2 (a) Self-association and (b) decomposition of the phenalenyl radical under atmospheric conditions.

For phenalenyl radicals, and in fact most carbon-centered radicals, delocalization is often not enough to overcome decomposition pathways. Such radicals therefore often require additional support in the form of steric protection or through the incorporation of heteroatoms into the π-system (*vide infra*). With respect to self-association of phenalenyl radicals, strategic placement of *tert*-butyl groups at the β-carbon atoms as in **1-9** (Chart 1.2) protects molecules from σ-dimerization both in solution and solid state.[28] However, the crystal structure reveals one-dimensional (1D) stacks of molecules in an anti-staggered arrangement separated by interplanar distances within the van der Waals radii for carbon atoms. Studies have elucidated this close contact to exist as a multicenter/two-electron interaction, illustrating another mode of spin pairing between π-systems known as π-dimerization, as highlighted in **1-10**.[29–31] Like σ-dimerization, the degree of this bonding interaction is heavily influenced by the interplay of attractive and repulsive forces between molecules.[32] Thus, dimerization can be modulated through judicious choice of substituents along the radical framework.[33–35] For instance, chlorination of the peripheral atoms of phenalenyl affords **1-11**, in which electronic repulsion between these atoms disrupts π-dimerization along a stack of radicals in the solid state.[36]

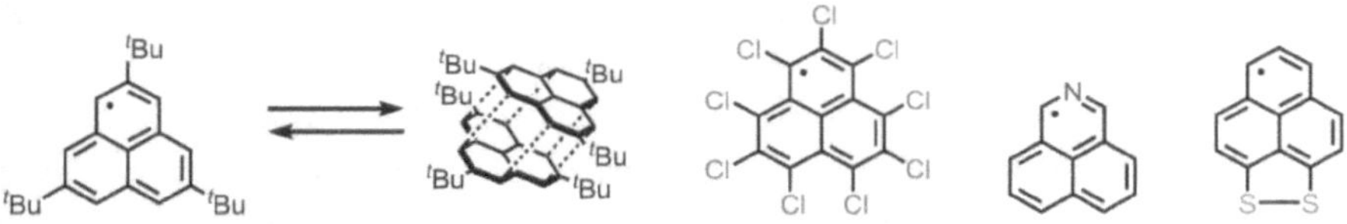

Chart 1.2 Sterically and electronically stabilized phenalenyl derivatives.

1.2.3 Stability Enhancement *via* Heteroatomic Effects

An alternative method employed to stabilize the unpaired electron in organic radicals takes advantage of heteroatoms that can contribute to the π-system such as nitrogen, oxygen and sulfur atoms. The energy of the SOMO can be further stabilized by the greater electronegativity of these atoms relative to carbon, while also promoting spin delocalization throughout the molecule. Commonly, C–H groups of carbon-based π-frameworks are replaced with an isolobal nitrogen as depicted in 2-azaphenalenyl **1-12**[37] or, as in the case of **1-13**, heteroatomic substituents can be included along the molecular edge (Chart 1.2).[38] Relative to unfunctionalized phenalenyl (i.e., **1-6**), the stability of **1-12** has only been investigated in solution where σ-dimerization is suppressed, while **1-13** has been shown to be electronically stabilized both in solution and solid state without steric support. The heteroatomic modifications to the phenalenyl skeleton, in addition to steric influences mentioned previously, certainly modulate the degree and type of dimerization between radicals within the various states; however, their ability to tolerate water and oxygen has yet to be reported.

4

Through solely an electronic approach, enhancing the stability of phenalenyl-based radicals has been achieved *via* extension of the molecular backbone into larger π-frameworks with the aid of heteroatom inclusion (see Chart 1.3 for examples). In **1-14**, the unpaired electron is delocalized by spiro-conjugation[39] between two orthogonal phenalenyl units connected through a tetrahedral boron center.[40] Depending on the hydrocarbon groups attached to the nitrogen atoms, derivatives of **1-14** have shown to be monomeric in crystalline form or assemble into an arrangement of π-motifs.[40–43] Preparation of extended phenalenyl frameworks while maintaining 3-fold symmetry continue to garner attention, an example of which is 4,8,12-trioxotriangulene **1-15**.[44] Here, radicals are electronically stabilized by a 25π-electron system and exhibit high resilience towards molecular oxygen in both solution and the solid state. Under aerobic conditions, the half-life of **1-15** concentrated in solution is 18.5 days, while solid samples possess high thermal stability up to 350°C.[44] It is noteworthy to mention other carbon-based radicals that display extensive π-delocalization sans phenalenyl, such as porphyrinoids. The structural diversity behind the macrocyclic framework presents a platform for stable open-shell structures utilizing varying ring sizes and substituents.[4,45] For example, the odd-alternant pentathiapentaphyrin **1-16** also exists as a planar 25π radical and has been demonstrated to be stable thermally and chemically towards air and moisture.[46]

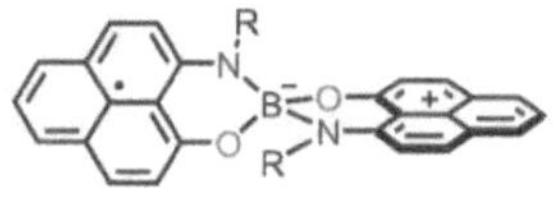

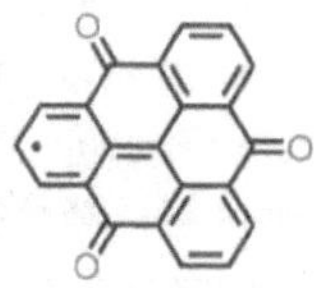

Chart 1.3 Examples of stable carbon-centered radicals with extended π-conjugation.

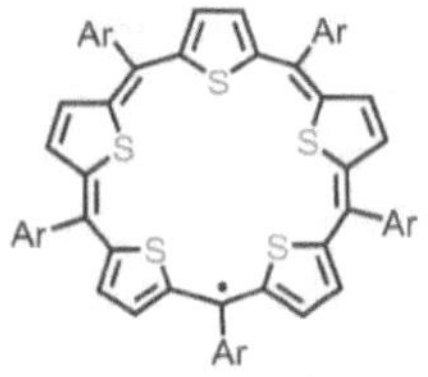

When spin density primarily resides on carbon atoms, as observed in the examples above, radicals benefit significantly from greater π-delocalization in order to circumvent reactivity. However, the same degree of resonance is not often required for heteroatom-centered radicals. For these radicals, σ-dimerization between light heteroatoms (e.g., N, O, S) is often impeded by the repulsive interactions between lone pairs, while reaping the electronic stabilization that accompanies the electronegativity.[47] This can be appreciated among some of the most stable and recognizable organic radicals shown in Chart 1.4. Nitroxide radicals including TEMPO (2,2,6,6-tetramethylpiperidin-1-yl)oxyl) and its analogues **1-17** are effective at stabilizing an unpaired electron by delocalization between the two heteroatoms,[48] and more so nitronyl nitroxide radicals **1-18**, in which the SOMO is partitioned between a pair of N–O moieties.[49] Generally speaking, these radicals display excellent resistance towards dimerization and can be handled in the presence of water and air, although they have the tendency to disproportionate in which substitution at the α-positions to the nitrogen atom is imperative. On the other hand, similar stabilities are afforded in verdazyls **1-19** and 6-oxoverdazyls **1-20**, many of which do not require the same extent of steric demand.[50–54] In these examples, two hydrazyl linkages are integrated into a cyclic arrangement, thereby optimizing delocalization of an unpaired electron between the moieties. This stability is not translated to most acyclic variants, whereby destabilizing effects arising from the twisting of the N–N bond induce a persistent nature.[55]

1-17   **1-18**   **1-19**   **1-20**

Chart 1.4 Family of nitroxides and verdazyls as stable nitrogen- and oxygen-centered π-radicals.

Molecular radicals that introduce heavier main group elements into the organic infrastructure are often thermodynamically unstable and require sufficient steric encumberment when spin density is centralized on these atoms.[56,57] These systems in many cases do not benefit from the electronic effects afforded by light heteroatoms and often suffer poor p-overlap for π-delocalization as a consequence of atomic size differences. To the contrary, an exception lies with sulfur-based radicals, more specifically, the family of thiazyl-based radicals bearing an unsaturated three-electron S–N linkage (i.e., π^* SOMO).[58–60] Incorporation of the thiazyl unit into five- and six-membered heterocyclic skeletons has resulted in a series of 7π-electron radicals (**1-21–1-25** in Chart 1.5).[60–64] In general, monocyclic thiazyls are persistent and require meticulous handling to avoid hydrolysis[65] and oxidation.[66] In the absence of moisture and air, the electronic effects associated with these atoms are enough to stabilize the open-shell structure without the need for steric interventions in solution. Moreover, the difference in electronegativity between the S–N bond leaves the moiety polarized towards the nitrogen atom. With the lack of protecting groups, this polarization has shown to be influential towards controlling the self-assembly of radicals through electrostatic $S^{\delta+}\cdots N^{\delta-}$ interactions, from which this family of radicals have portrayed a diverse array of crystallographic architectures; an indispensable quality for applications utilizing molecular radicals (*vide infra*). Despite this desirable attribute, the vast majority of monocyclic thiazyls undergo a degree of π-dimerization in the solid state.[67]

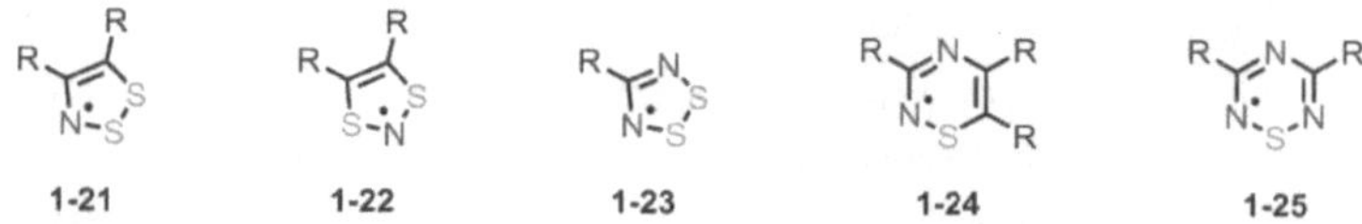

1-21   1-22   1-23   1-24   1-25

Chart 1.5 Family of five- and six-membered monocyclic thiazyl π-radicals.

Inhibition of dimerization through an electronic means has been demonstrated through the annulation of two monocylic radicals with a bridging system (Chart 1.6). For example, the unsaturated C–C bond adjacent to the spin-bearing nitrogen atom in **1-21** permits functionalization to a conjugated unit such as pyridine, thereby bridging the spin density from one 1,2,3,5-dithiadiazolyl ring between two rings as depicted in bisdithiazolyls **1-26**. The synthetic preparation for these molecules has thus far required alkyl functionalization of the pyridyl nitrogen atom with a range of substituents at the basal carbon atom (e.g., proto, halo, methyl, phenyl).[68,69] In the majority of cases, resonance stabilization contributes to discrete radicals of **1-26** in the solid state, from which the spin-bearing molecules have demonstrated unique packing arrangements arising from favourable intermolecular interactions (e.g., S$\cdots$N', π–π). Other variations of this framework have been explored to probe the structure-property relationships of **1-26**. For instance, **1-24** contains the structural infrastructure to be linked to bridging systems, as demonstrated by pyridine-bridged bisthiadiazinyls **1-27**. The few known derivatives all π-stack as unassociated radicals in the solid state.[70,71] Alternatively, modifying the bridging ring system in **1-26** to pyrazine or semiquinone as in **1-28** and **1-29**, respectively, influences the supramolecular architectures drastically. In **1-28**, the radicals are more susceptible to self-association as a consequence of the removal of the basal carbon unit,[72,73] while in **1-29**, replacement of pyridyl nitrogen for the isolobal carbonyl moiety induces close packing without such distortions.[74–77] Thus, the role of the exocyclic ligands towards modulating the stability is highlighted within this family of fused tricyclic radicals. Furthermore, replacement of the sulfur atoms for selenium has been successful in maintaining the stable radical character in the solid state (i.e., **1-30**).[78]

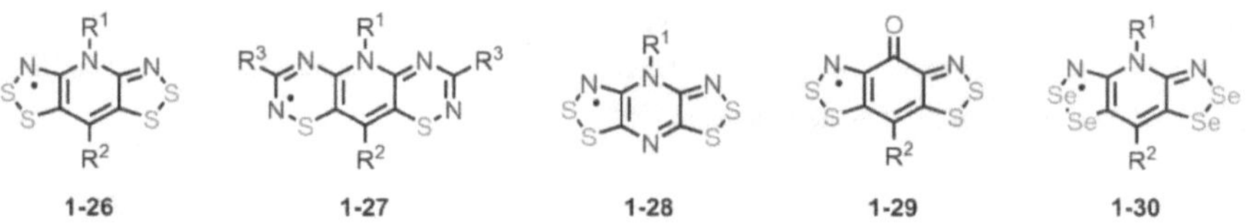

1-26   1-27   1-28   1-29   1-30

Chart 1.6 Family of bridged bisthiazyl and bisselenazyl π-radicals.

To date, numerous persistent and stable molecular radicals have been constructed using covalent design strategies, in which stabilization is afforded through a degree of steric, conjugative, or inductive effects, and for many, an interplay of all three. The enhancements made towards the lifetime of open-shell molecules has led to their widespread development in an array of chemical, biological and physical applications. Crucial to the design of these materials is a molecular framework that not only stabilizes the electronic structure but supports the properties of a targeted application. In that regard, radicals have been at the forefront as novel building blocks towards the design of molecular materials, where bulk properties can be achieved through the assembly of tailored molecules. The unpaired electron intrinsically brings a novel conductive and magnetic element to the material, that can be activated through pathways that enable effective communication between electron sites. Crystal engineering of the solid state becomes essential, such that favourable and predicable interactions can be integrated through the molecular design of the radical. Thus, there remains a growing interest to unlock new molecular templates that allow control over these properties with precision. As this book pertains to the design of molecular materials, a short discussion will follow on fundamental properties relating to open-shell systems, and more specifically, molecular radicals.

1.3 Electronic Properties of Molecular Radicals

1.3.1 Origin of Magnetism

The inherent magnetic properties of open-shell systems stem from both the charge and spin of an electron.[79] Spin is best understood as an intrinsic form of angular momentum.[80] Although there is no classical equivalent to this quantum phenomenon, one often conceptualizes the spin moment as a perpendicular vector $\vec{S}$ (Figure 1.1a, red arrow) arising from the rotation of an electron about a fixed axis. The quantization of $\vec{S}$ is defined by Equation 1.1 and parameterized by the spin quantum number s, in which for all electrons, $s = \frac{1}{2}$, and the reduced form of Planck's constant ($\hbar = h/2\pi$). Although this vector carries a direction in three-dimensional (3D) space, it cannot be measured directly as a contingency of the Heisenberg uncertainty principle, but instead a component of $\vec{S}$ can be measured along a given axis with precision. Quantum mechanics dictates that the magnitude of this projection S_z must take on specific values agreeing with Equation 1.2, assuming the direction of this axis to be z of a Cartesian coordinate system. In addition, S_z can assume one of two positions as depicted in Figure 1.1.b and therefore, the quantum number m_s replaces s to distinguish between the spin states coparallel ($m_s = +\frac{1}{2}$; spin up or α-spin) and antiparallel ($m_s = -\frac{1}{2}$; spin down or β-spin) to the z-direction.

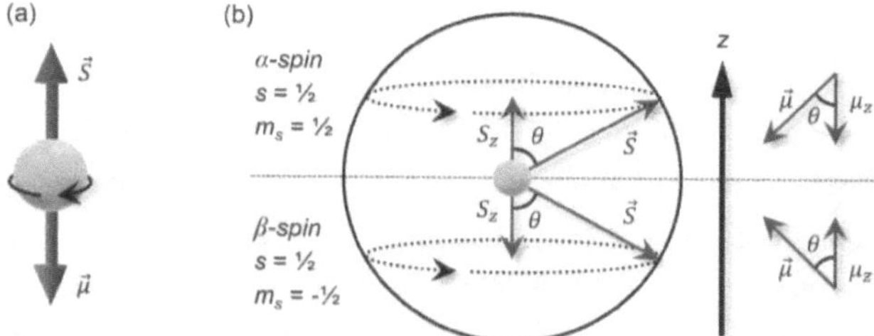

Figure 1.1 (a) Pictorial representations of the intrinsic spin $\vec{S}$ and magnetic moment $\vec{\mu}$ of an electron. (b) Projections of the vectors $\vec{S}$ and corresponding $\vec{\mu}$ along the z-direction for spin up (above dashed line) and spin down (below dashed lines) configurations as defined by m_s. Dotted arrows illustrate the precession of $\vec{S}$ around the direction of an external magnetic field $\vec{H}$ aligned with z.

$$\vec{S} = \hbar\sqrt{s(s+1)} \tag{1.1}$$

$$|S_z| = \hbar s = |\hbar m_s| \tag{1.2}$$

Consequentially, a magnetic field is generated as a result of a spinning particle bearing both mass m and charge q. The magnetic moment for an electron, signified by the vector $\vec{\mu}$ in Equation 1.3, is derived by the relationship between the mass and charge of the particle with its spin through a quantum-mechanical proportionality constant g. With respect to the angular momentum, the direction of $\vec{\mu}$ is antiparallel to $\vec{S}$, originating from the negative charge of an electron (Figure 1.1a, blue arrow) and subjected to same uncertainty. In other words, projections of $\vec{\mu}$ can only be measured along a given axis (μ_z in Figure 1.1b) with values determined by Equation 1.4, where the constant term ($\hbar q/2m$) is substituted with the Bohr magneton μ_B.

$$\vec{\mu} = -g\frac{q}{2m}\vec{S} \tag{1.3}$$

$$\mu_z = -g\frac{q}{2m}\hbar m_s = -g\mu_B m_s \tag{1.4}$$

Just like spin, orbital angular momentum $\vec{L}$ produces a second magnetic moment. These motions have the potential to be coupled through a spin-orbit interaction, in which case, the total electronic angular momentum $\vec{J}$ is the vector sum of the individual observables (i.e., $\vec{S} + \vec{L}$). Thus, the total magnetic moment of an electron along a given axis will reflect this summation. It is noteworthy that spin-orbit interactions are usually quenched in light atoms with principle quantum number of 2 and therefore, the predominate source for the total magnetic moment of an electron in these atoms is rooted from spin. Collectively, each electron contributes to the non-zero magnetic moment of a material that constitutes its magnetic behaviour.

1.3.2 Electron-Zeeman Interaction

In the presence of an external magnetic field $\vec{H}$, electrons will experience a torque arising from the interaction between its magnetic moment and the field. For unpaired electrons, this force induces a precession of $\vec{S}$ around the direction of $\vec{H}$ like a spinning top, such that the most stable orientation occurs when $\vec{S}$ is maintained at a conical angle of $\theta = 54.73°$. This angle coincides with maintaining the component S_z along this direction (dotted arrows in Figure 1.1b). Given the quantum-mechanical nature of spin, the potential energy of this system emerging from the magnetic interaction can be described by the spin Hamiltonian operator in Equation 1.5, such that the energy of each spin state E is quantized according to Equation 1.6.[81,82] Since both spin states differ in the direction, not magnitude of μ_z, the degeneracy of the energy states is broken by the magnetic field as to produce an energy gap ΔE_{EZ} outlined in Equation 1.7. This effect, known as the electron-Zeeman interaction, is illustrated for a one-spin system in Figure 1.2, in which the spin-down state of an electron is favoured over the other (i.e., $E_\beta < E_\alpha$).

$$\hat{\mathcal{H}}_{EZ} = -(\vec{\mu} \cdot \vec{H}) \tag{1.5}$$

$$E = -(-g\mu_B m_s \cdot H) \tag{1.6}$$

$$\Delta E_{EZ} = g\mu_B \hbar \cdot H \tag{1.7}$$

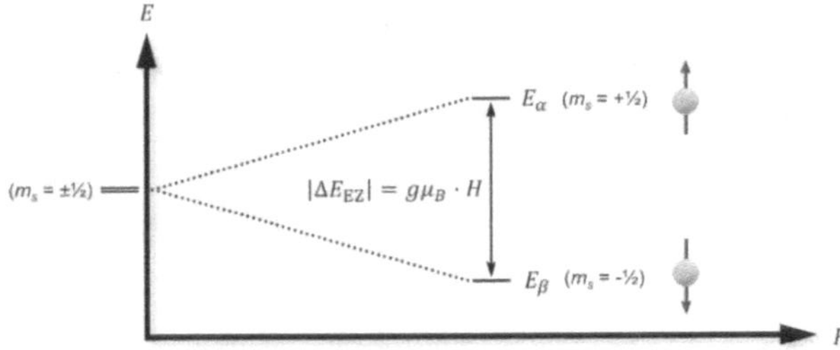

Figure 1.2 Zeeman energy splitting diagram of the m_s states for an electron as a function of the strength of an external magnetic field H.

From the Zeeman interaction, the spin state of a material can be manipulated for different purposes. Spin inversions can be facilitated through the application of electromagnetic radiation that correspond to ΔE_{EZ}. The radiative emissions upon the relaxation from excited spin states is the basis of electron paramagnetic resonance (EPR) spectroscopy.[83] Analogous to nuclear magnetic resonance, EPR can be applied to the identification of open-shell molecules,[84] characterization of electronic structures and imaging techniques that rely on spin labels.[7] Conversely, thermal energy can populate high-energy spin states. The changes in the magnetic behaviour of a material as a function of temperature is a focus of magnetometry.[79]

For isolated spins, the net magnetization of a material depends on a Boltzmann distribution of the spin state populations; however, in instances where magnetic communication between spin sites is present, an ordered alignment of spins can prevent thermal population to excited spin states. Upon removal of the field, a bulk magnet can be produced if the communication is strong enough and the retention of the spin alignment is achieved.

1.3.3 Molecular Magnetism and Conductivity

Magnetic communication of the alignment of spins originates from the exchange interaction between electrons. For two spin-bearing sites (i,j) in close proximity to each other, the following Heisenberg Hamiltonian in Equation 1.8 is used to define the coupling of spins by the exchange parameter J, where the interaction spin operators $\widehat{S}_i$ and $\widehat{S}_j$ can either be ferromagnetic (FM; parallel spins) or antiferromagnetic (AFM; antiparallel spins) in nature.[85] The spin quantum number S is used to reflect the total number of spins for a multi-electron system (i.e., $S = 1$), such that the spin multiplicity $(2S + 1)$ describes the possible spin configurations that can be achieved through values of m_s (i.e., $m_s = S, S - 1, \ldots -S$). In this bicentric system, the energies of the triplet and singlet spin-ground states will be determined by the magnitude of J, in which FM interactions favour positive values of J, while AFM interactions are determined by negative values. Thus, the difference in energy between the two configurations ΔE_{ST} is a function of $2J$ (Equation 1.9) as illustrated in Figure 1.3.

$$\widehat{\mathcal{H}}_{EZ} = -(\vec{\mu} \cdot \vec{H}) \tag{1.8}$$

$$E = -(-g\mu_B m_s \cdot H) \tag{1.9}$$

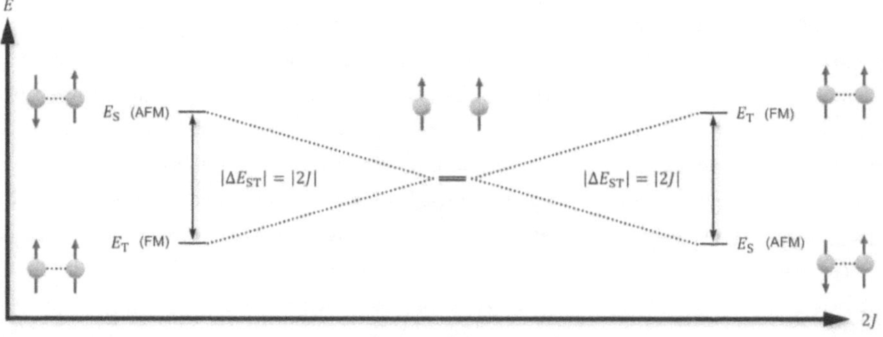

Figure 1.3 Energy diagram of a two-spin system as a function of the exchange integral J.

The mechanism for magnetic exchange is dependent on a pathway facilitating the interaction, in which for molecular radicals with localized spins, often entails the direct overlap of neighbouring SOMOs. Using the two-site Hubbard model, the degree and magnitude of the pairwise exchange between radicals can be represented by two terms in Equation 1.10. The first energy term denotes the potential exchange integral (K) contributing to the preferential alignment of coparallel spins given by Hund's principles. The second term describes the kinetic exchange between spins by means of the transfer integral (t) and the on-site Coulombic repulsion (U), which simultaneously stabilizes an AFM interaction. More often than not, J is AFM due to the direct relationship of t with the overlap of the magnetic orbitals ($|t| > 0$; Figure 1.4a,c). On the basis of orthogonal overlap ($|t| = 0$; Figure 1.4b), FM interactions are expected to dominant solely through the exchange integral, which always remains positive.

$$2J_{ij} = 2K_{ij} - \frac{4\left(t_{ij}\right)^2}{U} \tag{1.10}$$

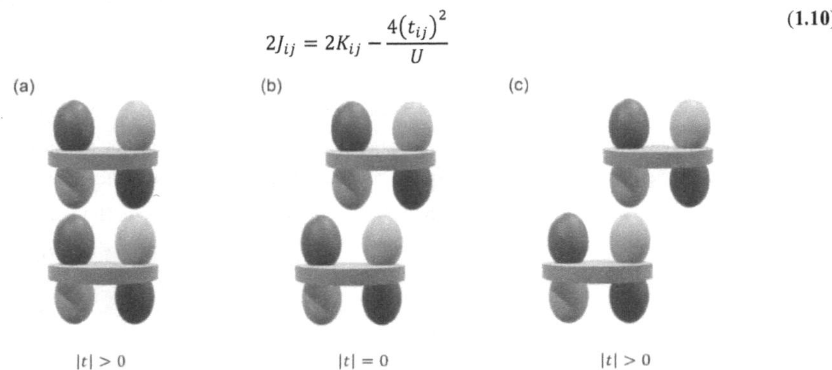

Figure 1.4 The magnitude of the transfer integral t influenced by the degree of SOMO overlap.

Given a lattice of interacting radicals, direct exchanges can construct the formation of one-, two- or 3D magnetic networks.[86] Assuming the latter option, exchange interactions can unify the spins of all radicals into a magnetically ordered state, where one can imagine a ferromagnet arising from the coparallel alignment of spins through a FM magnetic lattice, while the reverse would be true for an antiferromagnet. Characteristically, ferromagnets preserve magnetization upon the removal of a magnetic field and the stability of the lattice is measured by coercivity (i.e., the strength of a reverse field needed to demagnetize the system). Nevertheless, residual magnetization can be observed with AFM lattices. Examples include spin-canted antiferromagnets, where spins are not oriented uniformly with respect to each other, or ferrimagnets in which the magnitude of the magnetic moment differs between spin sites (e.g., $S_i \neq S_j$). For all magnetic orders, there is a critical temperature that corresponds to its transition (T_C for FM; T_N for AFM) that is dependent on the strength of J. Above this temperature, thermal energy is suffice to break apart these interactions, whereby higher-energy spin states can be populated and lead the system towards paramagnetic behaviour.

In addition to molecular magnetism, the Hubbard model can describe molecular conductivity in radicals (Figure 1.5).[87] From the previous discussion, the kinetic exchange between two spins sites was in relation to a high-spin state, where a large Columbic barrier (U) inhibits electron delocalization between the SOMOs. However, a low-spin state can be promoted through electron mobility if U is lower in energy in comparison to the term $4|t|$. The energy barrier to charge transfer can be approximated to the disproportionation enthalpy (ΔH_{disp}) shown in Equation 1.11, which corresponds to the energy difference between the ionization potential (IP) and electron affinity (EA) of a molecule. Molecular frameworks that can stabilize a neutral radical state and its respective oxidation states can effectively lower U and enhance conductivity within a system. Extension of a radical-dimer model to a 1D array of radicals generates the formation of half-filled molecular bands (Figure 1.5b), where the degree of bandwidth ($W = 4|t|$) with respect to U determines if a system will behave like a metal or an insulator. For metallic systems, conductivity is enhanced as electrons can, in essence, migrate freely from the valence band into the empty conduction band; however, this flow can be disrupted if the uniform π-stack of radicals is susceptible to a Peierls distortion.[88] In other words, the pairing of radicals provides greater stability to the system compared to the lattice interactions maintaining the uniform arrays, and thus, conductivity is compromised (Figure 1.5c).

$$U \approx \Delta H_{disp} = IP - EA \tag{1.11}$$

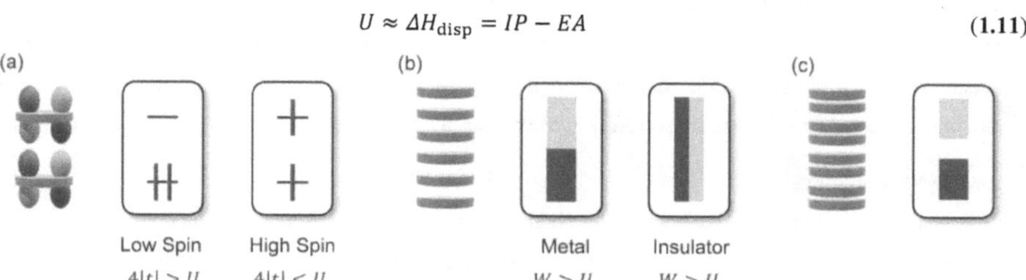

Figure 1.5 Extension of spin states from (a) a radical dimer to band structures for (b) a 1D array of SOMOS and (c) a Peierls distorted array.

1.4 Thiazyls as Building Blocks for Molecular Materials

1.4.1 Design of Molecular Materials

The design of radicals for molecular materials is dependent on several considerations.[9] Needless to say, a molecular framework must possess the ability to stabilize an open-shell structure in order to tap into the properties afforded from unpaired electrons. Stabilization of the radical character through a delocalized

π-system is beneficial from the perspective of intermolecular communication, as π–π interactions often dominate the assembly of molecules, while simultaneously creating opportunities for spin–spin interactions between radicals. Likewise, radicals profit from heteroatom inclusion in terms of their influences on electronic structures and their ability to direct crystal packing. This latter characteristic can be a vital tool for modulating multi-dimensional networks for electronic communication in molecular materials by influencing the degree and type of orbital overlap *via* intermolecular interactions. On the contrary, restricting intermolecular spin-spin interactions may be desirable, in which case, steric protection is not detrimental to the development of molecular materials. For example, isolation of spin centers necessitates the enhancement of physical properties, as seen for single-molecule magnets (SMMs) that rely heavily on magnetic anisotropy.[89,90]

With all this being said, an underlying factor for designing molecular materials is the ability to control the arrangement and orientation of molecules in the solid state rather than serendipitously. In the end, mother nature ultimately dictates this, however, the utility of supramolecular synthons can help generate predictable structural motifs.[91] A molecular platform that enables facile chemical modifications can provide a high degree of tunability in relation to its physicochemical properties. Ideally, control and modulation include both electronic and solid-state properties that comes through in the distribution of spin density, orbital energetics, and engineering of the crystal lattice, selectively. Out of the classes of radicals discussed this far, thiazyl radicals and their heavier chalcogen congeners are ideal candidates for molecular materials, as the structural versatility of these systems provide a systemic approach to fine-tuning the properties at hand. For this reason, the direction of the work presented in this book follows suit to understanding the structure-property relationships found in thiazyl molecules. With that, the focus of the remaining chapter will discuss an overview of molecular materials brought forth through thiazyls.

1.4.2 (SN)$_x$ Polymer

The pursuit of molecular materials utilizing thiazyl chemistry came to light upon the discovery of the (SN)$_x$ polymer (**1-31**); a one-component, metallic polymer synthesized exclusively from non-metal atoms.[92] This material crystallizes as fibrous needles that exhibits impressive conductive and superconductive properties.[93–96] The source of the metallic behaviour originates from unpaired electrons residing in π^* orbitals at each of the thiazyl subunits. Owing to the coparallel arrangement of planar chains (Figure 1.6), electrons can delocalize effectively along the polymeric backbone in a highly anisotropic manner.[97]

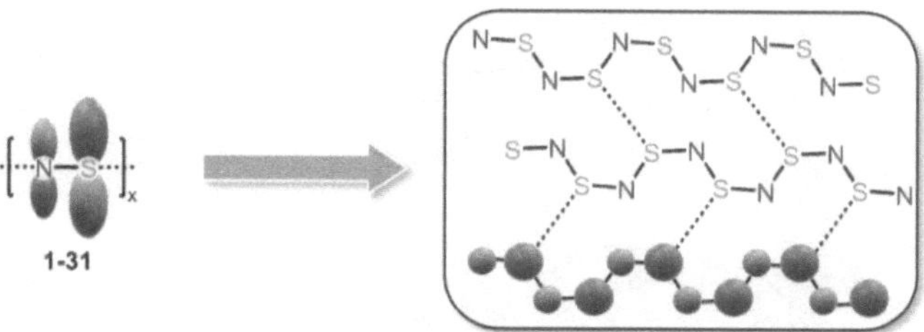

Figure 1.6 The metallic $(SN)_x$ polymer and its self-assembly into sheets through $S\cdots S'$ interactions (in box).

Inspired by **1-31**, thiazyl moieties have been integrated into a multitude of heterocyclic frameworks in the exploration for novel molecular materials.[60,98–100] The polarized S–N linkage has not only demonstrated to be a platform that stabilizes the spin and charge carrier but offers a means to direct crystal packing through predictable and favourable electrostatic interactions (i.e., $S\cdots N'$ and $S\cdots S'$). From this, a collection of supramolecular architectures has been realized through heterocyclic thiazyls, showcasing a range of multi-dimensional pathways for electronic communication. Bulk magnetic and conductive properties have been reported for many monomeric radicals in the solid state, and in some cases isostructural selenium variants have enhanced these properties, emphasizing the impact that larger, more diffuse orbitals have on strengthening interactions between spin-active molecules. Moreover, thiazyl molecules are well-equipped for the fine-tuning of the electronic and solid-state properties through the modifications of the skeletal framework and additionally has offered new opportunities to input additional physical features to the material (e.g., optical, liquid crystalline properties).

1.4.3 Monocyclic 1,2,3,5-Dithiadiazolyls as Molecular Magnets

1,2,3,5-Dithiadiazolyls (**1-23**) were one of the first thiazyl heterocyclic building blocks considered for the development of neutral radical conductors.[61,101–103] The attractive framework provides a local distribution of the π^*-SOMO along the intra-annular heteroatoms that features a nodal plane along the attachment of the exocyclic substituent (Figure 1.7). Because of this latter characteristic, functionalization of **1-23** at the carbon position has, for the most part, minimal effects on the electronic structure, but becomes influential on the crystallographic assembly. This valuable handle, in congruence with the antibonding symmetry of the SOMO, has been shown to manifest a diverse arrangement of π-stacking modes in the solid state as illustrated in Figure 1.7a-d.[67] Although, without steric support, there exists a proclivity for **1-23** to self-associate through the π-systems, given the dimerization enthalpy estimated to be *ca.* 35 kJ·mol⁻¹.

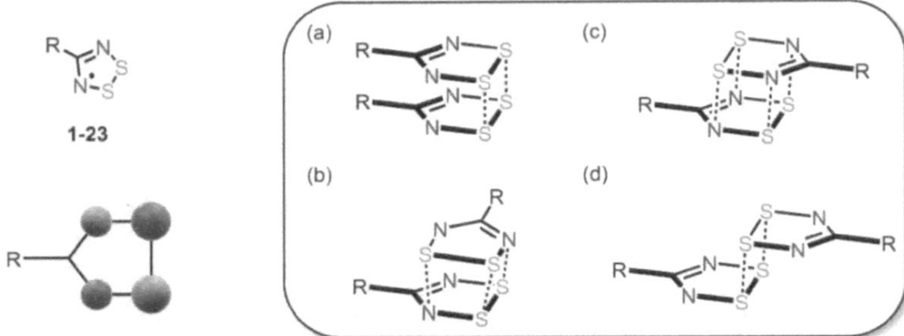

Figure 1.7 SOMO of 1,2,3,5-dithiadiazolyls **1-23** and (a) cis-cofacial (b) twisted (c) trans-cofacial (d) trans-antarafacial modes π-stacking modes in the solid state (in box).

With appropriate functionalization, molecules of **1-23** can remain disjointed in the solid state through steric and electronic repulsion. Examples of monomeric radicals listed in Figure 1.8 mark some of the first appearances of magnetic ordering seen in thiazyl-based systems. The aryl-substituted **1-32** can crystallize into either one of two polymorphs, in which case one exists as a bulk spin-canted antiferromagnet ordering below T_N = 36 K,[104,105] while the other precludes this long-range property.[106] In both structures, strategic placement of the *para*-nitrile group yields chain-like arrays of radicals linked head-to-tail by a series of CN···S' interactions, however, the striking differences in their magnetic behaviour lies within the orientation of chains with respect to each other (i.e., coparallel vs. antiparallel), that either propagates or disrupts a 3D magnetic lattice. When the nitrile synthon is replaced with a nitro group as in **1-33**, the system orders to a FM state below T_C = 1.3 K.[107] The sharp contrast in the magnetic properties is attributed to the out-of-plane twisting of the *para*-nitro synthon that continues to support a linear organization of radicals, but stabilizes a near-orthogonal arrangement of neighbouring chains.

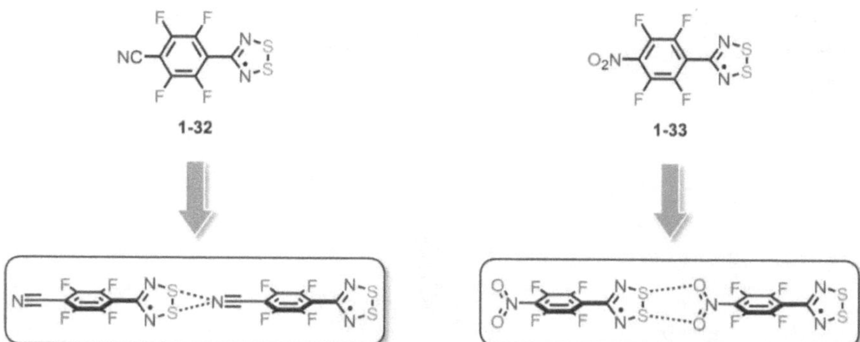

Figure 1.8 Examples of monomeric 1,2,3,5-dithiadiazolyls **1-32** and **1-33** and their assembly into chain-like arrays through interactions between their respective *para*-substituted synthon and disulfide linkage (in box).

1.4.4 Metal-Thiazyl Complexes Towards Magnetic Materials

While the exocyclic functional group in **1-23** plays a role in preventing self-association in the solid state, from a materials point of view this becomes counterproductive in strengthening spin-spin interactions through closer contacts. An alternative approach to circumvent these issues has explored the coordination of **1-23** to metal centers, which takes advantage of π-dimers reversibly dissociating in solution. Coordination complexes capable of stabilizing free radicals of **1-23** have been investigated for magnetic materials,[108,109] in hopes of facilitating an exchange pathway between paramagnetic metal centers *via* spin-bearing ligands.[110] As seen through the following bidentate ligands in Figure 1.9, radicals can be designed to chelate and bridge both transition and rare-earth metal ions in fixed geometries. The pyridine-functionalized ligand **1-34** provided a proof-of-concept of this approach for 1,2,3,5-dithiadiazolyls through chelation with Co(hfac)$_2$ (hfac = hexafluoroacetylacetonate) to afford the octahedral complex **1-35**.[111] On its own, the radical succumbs to π-dimerization in the solid state, although when complexed the radical character is shown to be stabilized and FM coupled to the high-spin Co^{II} center. Since that point, long-range magnetic ordering at low temperatures has been achieved through bridging ligands. Complexation of the diradical ligand **1-36** with Mn(hfac)$_2$ results in the bridged dinuclear complex **1-37**, in which a spin ground state of $S = 4$ arises through the AFM exchange interaction between the diradical ligand ($S = 1$) and Mn^{II} ions ($S = 5/2$ each).[112] Between neighbouring dinuclear complexes, electrostatic interactions proliferate a two-dimensional (2D) FM network through S $\cdots$ O' interactions arising from the ligands and weaker AF interactions couple these magnetic sublattices that order below 4.5 K. With respect to the bridging ligand

1-38, complexation with Sm(hfac)₃ has resulted in the formation of the one-dimensional coordination polymer **1-39** that magnetically orders below 3 K in a FM state.[113]

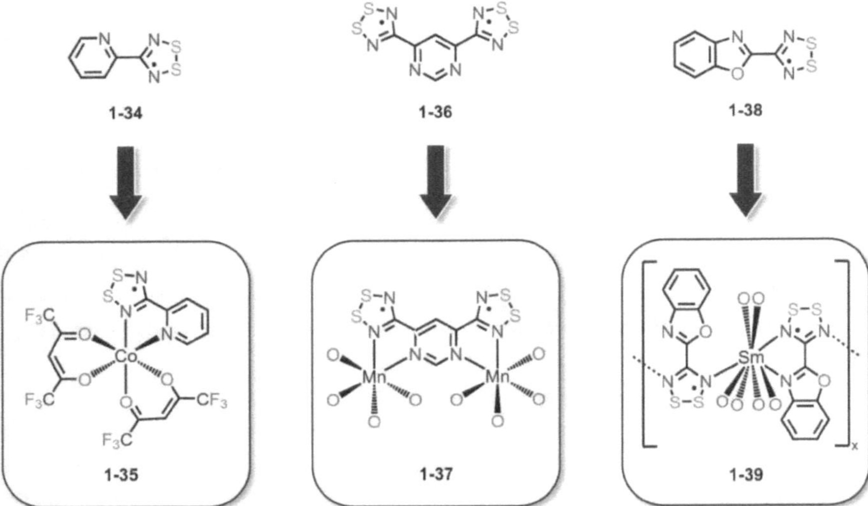

Figure 1.9 Examples of 1,2,3,5-dithiadiazolyls spin-bearing chelating ligands and coordinated complexes (in box). For clarity, only the oxygen atoms of the hexafluoroacetylacetonate ligands are drawn in complexes **1-37** and **1-39**.

Alongside **1-23**, 1,2,4,6-thiatriazinyls (**1-25**; Figure 1.10) were initially explored for the purpose of molecular conductivity.[114] These thiazyl frameworks also have the proclivity to dimerize along the N–S–N region in a cofacial manner, since the largest coefficient of the SOMO is located at the sulfur atom. In contrast to **1-23**, small coefficients lie on the carbon centers of the thiatriazinyl ring, in which case spin density can be influenced to an extent through the substituents at these positions.[115–117] With that in mind, the synthesis of heterocyclic-functionalized radicals were explored in my graduate studies towards the development of redox-versatile spin-active ligands. Work presented on these systems will be further discussed in the earlier chapters of this book.

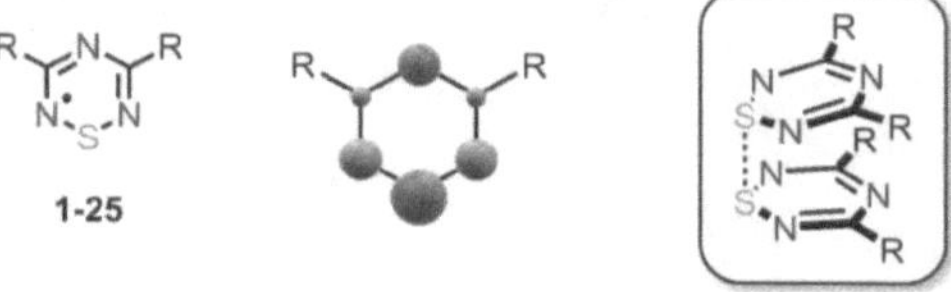

Figure 1.10 SOMO of 1,2,4,6-thiatriazinyls **1-25** and cis-cofacial mode of π-dimerization (in box).

1.4.5 Magneto-Structural Correlations in Resonance-Stabilized Chalcazyls

Pyridine-bridged bis-1,2,3-dithiazolyls (**1-26**) spawn a new generation of stable thiazyls radicals capable of remaining discrete in the solid state. The extensive resonance afforded by the *N*-alkyl pyridine bridge partitions spin density between two 1,2,3-dithiazole rings equivalently.[118] The SOMO of **1-26**, illustrated in Figure 1.11, has a nodal plane lying along the central axis. Therefore, the beltline substituents bear minimal influence towards the intrinsic properties of **1-26**, but aid in the prevention of radical self-association in the solid state. As a consequence, this family of radicals typically adopt slipped π-stacks that lock into a herringbone pattern.

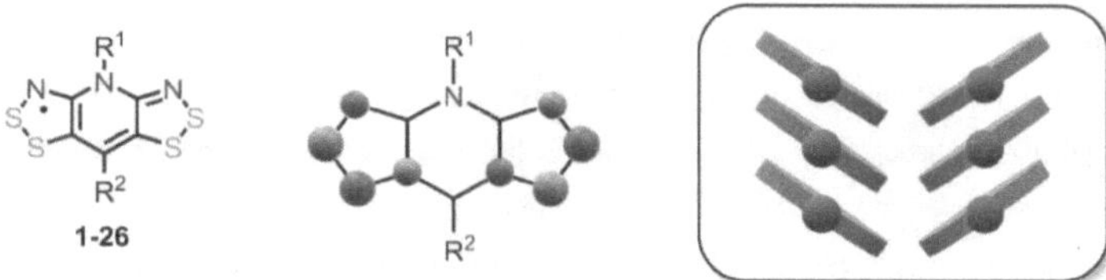

Figure 1.11 SOMO of 1,2,4,6-thiatriazinyls **1-25** and their slipped mode of π-stacking into herringbone arrangements (in box; radicals illustrated by grey disks with a blue sphere for the pyridyl nitrogen atom).

The relative sizes of the two exocyclic substituents has been used to probe the arrangement of π-stacks and, hence, the corresponding magnetic behaviour (Chart 1.7). Introduction of methyl groups as in **1-40** generates ribbon-like arrays of π-stacks connected by a single S$\cdots$S' contact.[119] The protruding alkyl groups pose a barrier for magnetic interactions between ribbons, thereby restricting magnetic dimensionality across the lattice. In the example of the *N*-ethylated analogue **1-41** with a proton at the basal carbon position, ribbon-like structures are woven together through additional S$\cdots$S' interactions in a response to the molecular modifications.[120] Concurrently, magnetic lattices are cross-linked in three dimensions, due to these sulfur atoms carrying the bulk of the spin density. Magnetic ordering was not observed in this derivative, however, replacement of sulfur for selenium led to isostructural variants showcasing a variety of magnetic anisotropic effects originating from greater spin-orbit coupling. Partial incorporation of selenium as for **1-42** orders to a spin-canted AFM state below $T_N = 18$ K.[121] Heavy atom effects are fully realized when complete selenium insertion is facilitated for **1-43**, in which the order temperature T_N is raised to 27 K.

1-40

1-41: $E^1 = E^2 = S$
1-42: $E^1 = S; E^2 = Se$
1-43: $E^1 = E^2 = Se$

Chart 1.7 Structural variants of the pyridine-bridged bisthiadiazinyl radical.

1.4.6 Pressure-Induced Conductivity *via* Resonance-Stabilized Chalcazyls

While the family of **1-26** and their selenium analogues exhibit rich magnetic behaviour, the lack of conductivity renders these systems insulating materials. The degree of slippage, in combination with steric influences, are contributing factors towards their poor performances as molecular conductors. Studies have shown enhancements to conductivity in **1-44** through the application of pressure (Chart 1.8).[122] At ambient pressure, the radicals FM order below 17.5 K and increases to 24 K at 2 GPa; one of the highest ordering temperatures for a non-metal ferromagnet. Further pressurization to 5 GPa results in the quenching of the magnetic behaviour, at which a weak metallic behaviour was observed at 9 GPa. The drastic change in electronic properties through applied pressure can be linked to the change in orthogonal orbital overlap by radical-radical slippage and the closer contacts enforced between them. Furthermore, the presence of selenium enhances the softness (reduction in U) and orbital overlap (promotion in t), which in turn, can improve their performance as molecular conductors.

1-44

1-45

Chart 1.8 Resonance-stabilized thiazyls exhibiting metallic states

Changing the bridging system to a semiquinone structure, as in **1-45** (Chart 1.8), has shown significant improvements on neutral radical conductivity.[123,124] A metallic state was reached at 3 GPa, owing to the removal of the *N*-alkyl group in place for an isoelectronic carbonyl, without the need for selenium incorporation. The chemical modification provides two benefits to the design of molecular conductors.[125,126] First, the carbonyl moiety helps to stabilize a 2D brick-like arrangement of radicals through favourable S···O' electrostatic interactions, from which an insulating state is destabilized by the lessening of the

Coulombic energy barrier. Secondly, although the carbonyl does not directly influence the energetics of the SOMO, it does contribute to the lowering of the LUMO such that it is nearly degenerate to the SOMO. This effectively amplifies the kinetic energy through the multiple orbital effect to help overcome the Columbic barrier to charge transfer at slightly higher pressures.

1.4.7 Magnetic Bistability in Thiazyl-Based Switches

Dimerization is prevalent amongst the majority of monocyclic thiazyls through a Peierls-type distortion along an evenly spaced π-stack of radicals. This π-bonding interaction generally weakens when the spin density on a thiazyls can be dispersed over a larger surface. For 1,3,2-dithiazolyls (**1-22**), adjacent carbon atoms permit the annulation of a conjugated ring system that can be used to drain some of the spin density located on the thiazyl ring. Many variants of **1-22** remain paramagnetic at room temperature and often self-associate through a solid-solid phase transition at lower temperatures. While dimerization is still viewed as an unfavourable attribute in these cases, transitions that are reversible and hysteretic open possibilities for switchable materials.[127] In other words, the monomer-dimer transition occurs at different temperatures upon thermal cycling, such that a bistable regime exists between these two points. For the present example, magnetic functionality can be inscribed from the differing spin-ground states of each crystallographic phase (i.e., $S = 0$ vs. $S = 1$).

Thermally induced magnetic bistability was first observed in **1-46** bearing a hysteresis window of 55 K centered around 150 K.[128] On two different accounts, removal of the central pyrazine bridge to afford **1-47** was shown to increase the range of the bistable regime and the transition temperatures, such that the switching capabilities were operational at room temperature.[129,130] A commonality between the crystallographic phases of **1-46** and **1-47** is the presence of slipped π-stacks of monomers and dimers in their respective high and low temperature structures (Figure 1.12). Crucial to the lagging behaviour is the lateral S$\cdots$N' and S$\cdots$S' interstack, proposed to stabilize each crystallographic phase.[131]

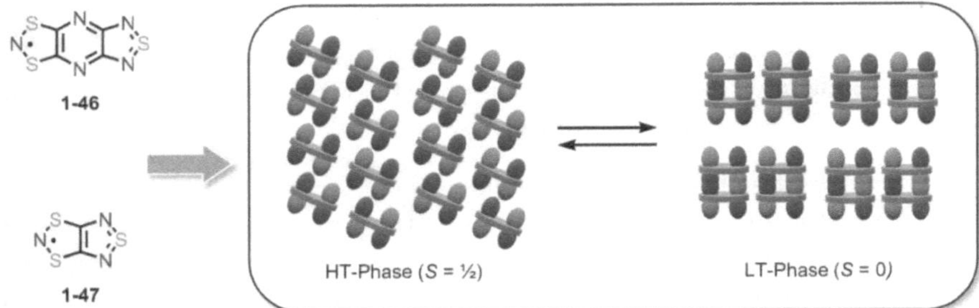

Figure 1.12 Magnetically bistable systems **1-46** and **1-47** exhibiting a paramagnetic high-temperature and a diamagnetic low-temperature phase.

1.4.8 Liquid Crystalline Materials *via* 1,2,4-Thiadiazinyl

Fused ring systems have been shown for the 1,2,4-thiadiazinyl radical framework (**1-24**) with two known crystal structures afforded by **1-48** and **1-49** (Chart 1.9).[132] Although the solid-state structures depict a Peierls-type distortion along the π-stacks of radicals, the intent of these systems was to be used for the development of liquid crystalline neutral radicals where the larger π-surface offers the rigid core for mesogenic materials. Precursors of **1-48** were functionalized with flexible arms, where **1-50** was shown to exhibit liquid crystalline properties and thus, the corresponding radical was envisioned to reciprocate a similar behaviour upon oxidation.[133] However, their sensitivity to oxygen was an underlying problem towards their fabrication. Such studies shed light for new prospective properties for other stable thiazyl variants.

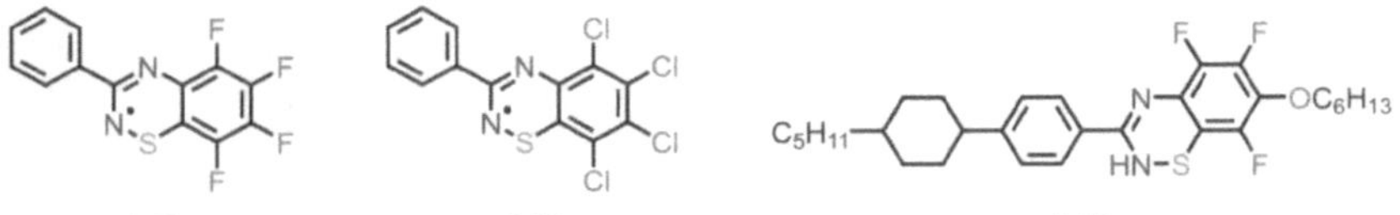

Chart 1.9 Benzannulated derivatives of 1,2,4-thiadiazinyls pursued for liquid crystalline materials.

1.5 book Outline

Herein, the breadth of the forthcoming work encompasses a body of work dedicated to the design of thiazyl-based solid state molecular materials. Research in the Brusso group has explored the bottom-up approach towards molecular materials utilizing both closed- and open-shell thiazyl building blocks (Chart 1.10). Early work focused on the structure-property relationship regarding the eight membered 1,5-dithia-2.4.6.8-tetrazocines (**1-51**) through thienyl functionalization. With the chemistry of these systems being relatable to open-shell thiazyls, we sought out to explore the influences of heteroaromatic functionalization of thiatriazinyls including the pyridyl- (**1-52**) and thienyl-functionalized (**1-53**) thiatriazinyls as described in Chapter 2. Results from this work illustrated the novelty of **1-52** and its reduced precursor **1-54** as a redox-versatile ligand system and directed us to explore molecular materials *via* coordination chemistry, from which Chapter 3 embarks on lanthanide chemistry utilizing **1-54** and provides insights on the self-assembly of dinuclear complexes. Furthermore, we recently expanded our work of thienyl-functionalization in bisthiadiazinyls as the resonance-stabilized framework is known to remain undimerized in the solid state. Unpublished work described in Chapter 4 surrounds the synthesis and challenges associated with the polymorphic nature of **1-55** and juxtaposed with its halogenated analogues **1-56** and **1-57**. Lastly, Chapter 5 entails the separation and isolation of two crystallographic phases of **1-58**, a bisdithiazolyl radical with mix-matched sizes of the beltline substituents, in which a highly symmetrical phase acquired at low temperatures illustrates the foundation for magnetic spiral lattices in the solid state.

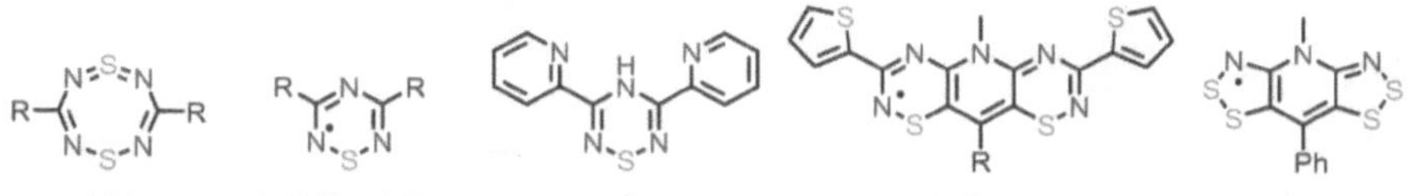

1-51

1-52 (R = 2-Py)
1-53 (R = 2-Th)

1-54

1-55 (R = H)
1-56 (R = F)
1-57 (R = Cl)

1-58

Chart 1.10 Open- and closed-shell building blocks studied in the Brusso group.

1.6 References

1   I. Fleming, *Molecular Orbitals and Organic Chemical Reactions*, John Wiley & Sons, Ltd, Chichester, UK, 2010.

2   P. Renaud and M. P. Sibi, *Radicals in Organic Synthesis*, Wiley, 2001.

3   R. G. Hicks, *Stable Radicals*, John Wiley & Sons, Ltd, Chichester, UK, 2010.

4   K. Kato and A. Osuka, *Angew. Chemie*, 2019, **131**, 9074–9082.

5   B. Tang, J. Zhao, J. F. Xu and X. Zhang, *Chem. Sci.*, 2020, **11**, 1192–1204.

6   D. Leifert and A. Studer, *Angew. Chemie Int. Ed.*, 2020, **59**, 74–108.

7   O. H. Griffith and A. S. Waggoner, *Acc. Chem. Res.*, 1969, **2**, 17–24.

8   J. Stubbe and W. A. van der Donk, *Chem. Rev.*, 1998, **98**, 2661–2662.

9   I. Ratera and J. Veciana, *Chem. Soc. Rev.*, 2012, **41**, 303–349.

10   A. J. Ihde, *Pure Appl. Chem.*, 1967, **15**, 1–14.

11   M. Gomberg, *J. Am. Chem. Soc.*, 1900, **22**, 757–771.

12   D. Griller and K. U. Ingold, *Acc. Chem. Res.*, 1976, **9**, 13–19.

13   H. Lankamp, W. T. Nauta and C. MacLean, *Tetrahedron Lett.*, 1968, **9**, 249–254.

14   W. P. Neumann, W. Uzick and A. K. Zarkadis, *J. Am. Chem. Soc.*, 1986, **108**, 3762–3770.

15   C. Rüchardt and H. -D Beckhaus, *Angew. Chemie Int. Ed. English*, 1980, **19**, 429–440.

16   M. Gomberg, *Berichte der Dtsch. Chem. Gesellschaft*, 1904, **37**, 1626–1644.

17   D. Dünnebacke, W. P. Neumann, A. Penenory and U. Stewen, *Chem. Ber.*, 1989, **122**, 533–535.

18   M. Ballester, J. Riera, J. Castafier, C. Badía and J. M. Monsó, *J. Am. Chem. Soc.*, 1971, **93**, 2215–2225.

19   M. Ballester, *Acc. Chem. Res.*, 1985, **18**, 380–387.

20   J. Veciana, J. Carilla, C. Miravitlles and E. Molins, *J. Chem. Soc. Chem. Commun.*, 1987, 812–814.

21   J. Guasch, X. Fontrodona, I. Ratera, C. Rovira and J. Veciana, *Acta Crystallogr. Sect. C Cryst. Struct. Commun.*, 2013, **69**, 255–257.

22   J. Veciana and M. I. Crespo, *Angew. Chemie Int. Ed. English*, 1991, **30**, 74–76.

23   D. H. Reid, *Tetrahedron*, 1958, **3**, 339–352.

24   D. H. Reid, *Q. Rev. Chem. Soc.*, 1965, **19**, 274.

25   K. Uchida and T. Kubo, *J. Synth. Org. Chem. Japan*, 2016, **74**, 1069–1077.

26   F. Gerson, *Helv. Chim. Acta*, 1966, **49**, 1463–1467.

27   K. Uchida, S. Ito, M. Nakano, M. Abe and T. Kubo, *J. Am. Chem. Soc.*, 2016, **138**, 2399–2410.

28   K. Goto, T. Kubo, K. Yamamoto, K. Nakasuji, K. Sato, D. Shiomi, T. Takui, M. Kubota, T. Kobayashi, K. Yakusi and J. Ouyang, *J. Am. Chem. Soc.*, 1999, **121**, 1619–1620.

29   J. S. Miller, *Acc. Chem. Res.*, 2007, **40**, 189–196.

30   S. Suzuki, Y. Morita, K. Fukui, K. Sato, D. Shiomi, T. Takui and K. Nakasuji, *J. Am. Chem. Soc.*, 2006, **128**, 2530–2531.

31   F. Mota, J. S. Miller and J. J. Novoa, *J. Am. Chem. Soc.*, 2009, **131**, 7699–7707.

32   D. Small, V. Zaitsev, Y. Jung, S. V. Rosokha, M. Head-Gordon and J. K. Kochi, *J. Am. Chem. Soc.*, 2004, **126**, 13850–13858.

33   V. Zaitsev, S. V. Rosokha, M. Head-Gordon and J. K. Kochi, *J. Org. Chem.*, 2006, **71**, 520–526.

34   Z. Mou, K. Uchida, T. Kubo and M. Kertesz, *J. Am. Chem. Soc.*, 2014, **136**, 18009–18022.

35   K. Uchida, Z. Mou, M. Kertesz and T. Kubo, *J. Am. Chem. Soc.*, 2016, **138**, 4665–4672.

36   P. A. Koutentis, Y. Chen, Y. Cao, T. P. Best, M. E. Itkis, L. Beer, R. T. Oakley, A. W. Cordes, C. P. Brock and R. C. Haddon, *J. Am. Chem. Soc.*, 2001, **123**, 3864–3871.

37   S. Zheng, J. Lan, S. I. Khan and Y. Rubin, *J. Am. Chem. Soc.*, 2003, **125**, 5786–5791.

38   L. Beer, S. K. Mandal, R. W. Reed, R. T. Oakley, F. S. Tham, B. Donnadieu and R. C. Haddon, *Cryst. Growth Des.*, 2007, **7**, 802–809.

39   R. Hoffmann, *Acc. Chem. Res.*, 1971, **4**, 1–9.

40   X. Chi, M. E. Itkis, B. O. Patrick, T. M. Barclay, R. W. Reed, R. T. Oakley, A. W. Cordes and R. C. Haddon, *J. Am. Chem. Soc.*, 1999, **121**, 10395–10402.

41   X. Chi, M. E. Itkis, K. Kirschbaum, A. A. Pinkerton, R. T. Oakley, A. W. Cordes and R. C. Haddon, *J. Am. Chem. Soc.*, 2001, **123**, 4041–4048.

42   S. K. Pal, M. E. Itkis, R. W. Reed, R. T. Oakley, A. W. Cordes, F. S. Tham, T. Siegrist and R. C. Haddon, *J. Am. Chem. Soc.*, 2004, **126**, 1478–1484.

43   S. K. Pal, M. E. Itkis, F. S. Tham, R. W. Reed, R. T. Oakley and R. C. Haddon, *Science*,

2005, **309**, 281–284.

44   Y. Morita, T. Murata, A. Ueda, C. Yamada, Y. Kanzaki, D. Shiomi, K. Sato and T. Takui, *Bull. Chem. Soc. Jpn.*, 2018, **91**, 922–931.

45   D. Shimizu and A. Osuka, *Chem. Sci.*, 2018, **9**, 1408–1423.

46   T. Y. Gopalakrishna, J. S. Reddy and V. G. Anand, *Angew. Chemie Int. Ed.*, 2014, **53**, 10984–10987.

47   C. A. Coulson, *Valence*, Oxford University Press, London, 2nd edn., 1961.

48   H. Karoui, F. Le Moigne, O. Ouari and P. Tordo, in *Stable Radicals: Fundamentals and Applied Aspects of Odd-Electron Compounds*, ed. R. G. Hicks, John Wiley & Sons, Ltd, Chichester, UK, 2010, pp. 173–229.

49   E. F. Ullman, J. H. Osiecki, D. G. B. Boocock and R. Darcy, *J. Am. Chem. Soc.*, 1972, **94**, 7049–7059.

50   R. Kuhn and H. Trischmann, *Angew. Chemie Int. Ed. English*, 1963, **2**, 155–155.

51   F. A. Neugebauer and H. Fischer, *Angew. Chemie Int. Ed. English*, 1980, **19**, 724–725.

52   E. C. Paré, D. J. R. Brook, A. Brieger, M. Badik and M. Schinke, *Org. Biomol. Chem.*, 2005, **3**, 4258–4261.

53   R. G. Hicks, *Org. Biomol. Chem.*, 2007, **5**, 1321–1338.

54   R. G. Hicks, in *Stable Radicals*, John Wiley & Sons, Ltd, Chichester, UK, 2010, pp. 245–279.

55   F. A. Neugebauer, *Angew. Chemie Int. Ed. English*, 1973, **12**, 455–464.

56   P. P. Power, *Chem. Rev.*, 2003, **103**, 789–809.

57   T. Chivers and J. Konu, in *Comprehensive Inorganic Chemistry II*, eds. J. Reedijk and K. Poeppelmeier, Elsevier, 2013, vol. 1, pp. 349–373.

58   T. Chivers, *Chem. Rev.*, 1985, **85**, 341–365.

59   R. T. Boeré and T. L. Roemmele, in *Comprehensive Inorganic Chemistry II*, eds. J. Reedijk

and K. Poeppelmeier, Elsevier, Elsevier., 2013, vol. 1, pp. 375–411.

60   R. G. Hicks, in *Stable Radicals: Fundamentals and Applied Aspects of Odd-Electron Compounds*, John Wiley & Sons, Ltd, Chichester, UK, 2010, pp. 317–380.

61   J. M. Rawson, A. J. Banister and I. Lavender, in *Advances in Heterocyclic Chemistry*, ed. A. R. Katritzky, Elsevier, 1995, vol. 62, pp. 137–247.

62   R. T. Oakley, in *Progress in Inorganic Chemistry*, ed. S. J. Lippard, John Wiley & Sons, Ltd, 2007, vol. 36, pp. 299–391.

63   O. A. Rakitin, *Russ. Chem. Rev.*, 2011, **80**, 647–659.

64   C. P. Constantinides and P. A. Koutentis, in *Advances in Heterocyclic Chemistry*, eds. C. A. Ramsden and E. F. V. Scriven, Elsevier, 2016, vol. 119, pp. 173–207.

65   L. D. Huestis, M. L. Walsh and N. Hahn, *J. Org. Chem.*, 1965, **30**, 2763–2766.

66   A. W. Cordes and R. T. Oakley, *Acta Crystallogr. Sect. C*, 1987, **43**, 1645–1646.

67   K. E. Preuss, *Polyhedron*, 2014, **79**, 1–15.

68   L. Beer, R. W. Reed, J. L. Brusso, A. W. Cordes, R. C. Haddon, M. E. Itkis, K. Kirschbaum, D. S. MacGregor, R. T. Oakley and A. A. Pinkerton, *J. Am. Chem. Soc.*, 2002, **124**, 9498–9509.

69   K. Lekin, J. W. L. Wong, S. M. Winter, A. Mailman, P. A. Dube and R. T. Oakley, *Inorg. Chem.*, 2013, **52**, 2188–2198.

70   L. Beer, J. L. Brusso, R. C. Haddon, M. E. Itkis, H. Kleinke, A. A. Leitch, R. T. Oakley, R. W. Reed, J. F. Richardson, R. A. Secco and X. Yu, *J. Am. Chem. Soc.*, 2005, **127**, 18159–18170.

71   A. A. Leitch, R. T. Oakley, R. W. Reed and L. K. Thompson, *Inorg. Chem.*, 2007, **46**, 6261–6270.

72   A. A. Leitch, C. E. McKenzie, R. T. Oakley, R. W. Reed, J. F. Richardson and L. D. Sawyer, *Chem. Commun.*, 2006, 1088.

73   A. A. Leitch, R. W. Reed, C. M. Robertson, J. F. Britten, X. Yu, R. A. Secco and R. T.

Oakley, *J. Am. Chem. Soc.*, 2007, **129**, 7903–7914.

74   X. Yu, A. Mailman, P. A. Dube, A. Assoud and R. T. Oakley, *Chem. Commun.*, 2011, **47**, 4655–4657.

75   X. Yu, A. Mailman, K. Lekin, A. Assoud, C. M. Robertson, B. C. Noll, C. F. Campana, J. A. K. Howard, P. A. Dube and R. T. Oakley, *J. Am. Chem. Soc.*, 2012, **134**, 2264–2275.

76   X. Yu, A. Mailman, K. Lekin, A. Assoud, P. A. Dube and R. T. Oakley, *Cryst. Growth Des.*, 2012, **12**, 2485–2494.

77   J. W. L. Wong, A. Mailman, S. M. Winter, C. M. Robertson, R. J. Holmberg, M. Murugesu, P. A. Dube and R. T. Oakley, *Chem. Commun.*, 2014, **50**, 785–787.

78   J. L. Brusso, K. Cvrkalj, A. A. Leitch, R. T. Oakley, R. W. Reed and C. M. Robertson, *J. Am. Chem. Soc.*, 2006, **128**, 15080–15081.

79   J. M. D. Coey, *Magnetism and magnetic materials*, Cambridge University Press, 2010.

80   P. A. M. Dirac, *The Principles of Quantum Mechanics*, Oxford University Press, 4th edn., 1958.

81   W. Pauli, *Zeitschrift für Phys.*, 1925, **31**, 765–783.

82   P. A. M. Dirac, *Proc. R. Soc. London. Ser. A, Contain. Pap. a Math. Phys. Character*, 1928, **117**, 610–624.

83   F. Gerson and W. Huber, *Electron Spin Resonance Spectroscopy of Organic Radicals*, Wiley-VCH Verlag GmbH & Co. KGaA, Weinheim, 2003.

84   J. C. Walton, in *Encyclopedia of Radicals in Chemistry, Biology and Materials*, eds. C. Chatgilialoglu and A. Studer, John Wiley & Sons, Ltd, Chichester, UK, 2012.

85   W. Heisenberg, *Zeitschrift für Phys.*, 1926, **38**, 411–426.

86   S. M. Winter, S. Hill and R. T. Oakley, *J. Am. Chem. Soc.*, 2015, **137**, 3720–3730.

87   R. C. Haddon, *Nature*, 1975, **256**, 394–396.

88   R. E. Peierls, *Quantum Theory of Solids*, Oxford University Press, 2007, vol. 15.

89   G. Christou, D. Gatteschi, D. N. Hendrickson and R. Sessoli, *MRS Bull.*, 2000, **25**, 66–71.

90   E. Coronado, *Nat. Rev. Mater.*, 2020, **5**, 87–104.

91   G. R. Desiraju, *J. Am. Chem. Soc.*, 2013, **135**, 9952–9967.

92   A. J. Banister and I. B. Gorrell, *Adv. Mater.*, 1998, **10**, 1415–1429.

93   M. Goehring and D. Voigt, *Naturwissenschaften*, 1953, **40**, 482.

94   V. V Walatka, M. M. Labes and J. H. Perlstein, *Phys. Rev. Lett.*, 1973, **31**, 1139–1142.

95   R. L. Greene, G. B. Street and L. J. Suter, *Phys. Rev. Lett.*, 1975, **34**, 577–579.

96   M. H. Whangbo, R. Hoffman and R. B. Woodward, *Proc. R. Soc. Lond. A.*, 1979, **366**, 23–46.

97   K. Kaneto, M. Yamamoto, K. Yoshino and Y. Inuishi, *J. Phys. Soc. Japan*, 1979, **47**, 167–175.

98   J. M. Rawson, A. Alberola and A. Whalley, *J. Mater. Chem.*, 2006, **16**, 2560.

99   A. W. Cordes, R. C. Haddon and R. T. Oakley, *Phosphorus. Sulfur. Silicon Relat. Elem.*, 2004, **179**, 673–684.

100   J. M. Rawson and C. P. Constantinides, in *Materials and Energy*, ed. J. S. Miller, World Scientific Publishing Co. Pte Ltd, 2018, vol. 4, pp. 95–124.

101   A. W. Cordes, R. C. Haddon and R. T. Oakley, *Adv. Mater.*, 1994, **6**, 798–802.

102   D. A. Haynes, *CrystEngComm*, 2011, **13**, 4793–4805.

103   M. A. Nascimento and J. M. Rawson, in *Encyclopedia of Inorganic and Bioinorganic Chemistry*, John Wiley & Sons, Ltd., 2019, pp. 1–11.

104   A. J. Banister, N. Bricklebank, I. Lavender, J. M. Rawson, C. I. Gregory, B. K. Tanner, W. Clegg, M. R. J. Elsegood and F. Palacio, *Angew. Chemie Int. Ed. English*, 1996, **35**, 2533–2535.

105   M. Deumal, J. M. Rawson, A. E. Goeta, J. A. K. Howard, R. C. B. Copley, M. A. Robb and J. J. Novoa, *Chem. - A Eur. J.*, 2010, **16**, 2741–2750.

106   A. J. Banister, N. Bricklebank, W. Clegg, M. R. J. Elsegood, C. I. Gregory, I. Lavender, J. M. Rawson and B. K. Tanner, *J. Chem. Soc. Chem. Commun.*, 1995, 679.

107   A. Alberola, R. J. Less, C. M. Pask, J. M. Rawson, F. Palacio, P. Oliete, C. Paulsen, A. Yamaguchi, R. D. Farley and D. M. Murphy, *Angew. Chemie Int. Ed.*, 2003, **42**, 4782–4785.

108   K. E. Preuss, *J. Chem. Soc. Dalt. Trans.*, 2007, 2357–2369.

109   K. E. Preuss, *Coord. Chem. Rev.*, 2015, **289–290**, 49–61.

110   A. Caneschi, D. Gatteschi, R. Sessoli and P. Rey, *Acc. Chem. Res.*, 1989, **22**, 392–398.

111   N. G. R. Hearns, K. E. Preuss, J. F. Richardson and S. Bin-Salamon, *J. Am. Chem. Soc.*, 2004, **126**, 9942–9943.

112   E. M. Fatila, R. C. Clérac, M. Rouziè, ‖ Dmitriy, V. Soldatov, M. Jennings and K. E. Preuss, *J. Am. Chem. Soc*, 2013, **29**, 54.

113   E. M. Fatila, A. C. Maahs, M. B. Mills, M. Rouzières, D. V. Soldatov, R. Clérac and K. E. Preuss, *Chem. Commun.*, 2016, **52**, 5414–5417.

114   P. J. Hayes, R. T. Oakley, A. W. Cordes and W. T. Pennington, *J. Am. Chem. Soc.*, 1985, **107**, 1346–1351.

115   R. T. Boere, A. W. Cordes, P. J. Hayes, R. T. Oakley, R. W. Reed and W. T. Pennington, *Inorg. Chem.*, 1986, **25**, 2445–2450.

116   R. T. Boeré and T. L. Roemmele, in *Phosphorus, Sulfur and Silicon and the Related Elements*, 2004, vol. 179, pp. 875–882.

117   R. T. Boeré, T. L. Roemmele and X. Yu, *Inorg. Chem.*, 2011, **50**, 5123–5136.

118   L. Beer, R. W. Reed, J. L. Brusso, A. W. Cordes, R. C. Haddon, M. E. Itkis, K. Kirschbaum, D. S. MacGregor, R. T. Oakley and A. A. Pinkerton, *J. Am. Chem. Soc.*, 2002, **124**, 9498–9509.

119   L. Beer, J. F. Britten, O. P. Clements, R. C. Haddon, M. E. Itkis, K. M. Matkovich, R. T. Oakley and R. W. Reed, *Chem. Mater.*, 2004, **16**, 1564–1572.

120   L. Beer, J. F. Britten, J. L. Brusso, A. W. Cordes, R. C. Haddon, M. E. Itkis, D. S.

MacGregor, R. T. Oakley, R. W. Reed and C. M. Robertson, *J. Am. Chem. Soc.*, 2003, **125**, 14394–14403.

121   A. A. Leitch, J. L. Brusso, K. Cvrkalj, R. W. Reed, C. M. Robertson, P. A. Dube and R. T. Oakley, *Chem. Commun.*, 2007, 3368.

122   A. A. Leitch, K. Lekin, S. M. Winter, L. E. Downie, H. Tsuruda, J. S. Tse, M. Mito, S. Desgreniers, P. A. Dube, S. Zhang, Q. Liu, C. Jin, Y. Ohishi and R. T. Oakley, *J. Am. Chem. Soc.*, 2011, **133**, 6051–6060.

123   A. Mailman, S. M. Winter, X. Yu, C. M. Robertson, W. Yong, J. S. Tse, R. A. Secco, Z. Liu, P. A. Dube, J. A. K. Howard and R. T. Oakley, *J. Am. Chem. Soc.*, 2012, **134**, 9886–9889.

124   S. M. Winter, A. Mailman, R. T. Oakley, K. Thirunavukkuarasu, S. Hill, D. E. Graf, S. W. Tozer, J. S. Tse, M. Mito and H. Yamaguchi, *Phys. Rev. B*, 2014, **89**, 214403.

125   D. Tian, S. M. Winter, A. Mailman, J. W. L. Wong, W. Yong, H. Yamaguchi, Y. Jia, J. S. Tse, S. Desgreniers, R. A. Secco, S. R. Julian, C. Jin, M. Mito, Y. Ohishi and R. T. Oakley, *J. Am. Chem. Soc.*, 2015, 137, 14136–14148.

126   A. Mailman, C. M. Robertson, S. M. Winter, P. A. Dube and R. T. Oakley, *Inorg. Chem.*, 2019, **58**, 6495–6506.

127   J. M. Rawson, A. Alberola and A. Whalley, *J. Mater. Chem.*, 2006, **16**, 2560–2575.

128   T. M. Barclay, A. W. Cordes, N. A. George, R. C. Haddon, M. E. Itkis, M. S. Mashuta, R. T. Oakley, G. W. Patenaude, R. W. Reed, J. F. Richardson and H. Zhang, *J. Am. Chem. Soc.*, 1998, **120**, 352–360.

129   W. Fujita and K. Awaga, *Science*, 1999, **286**, 261–262.

130   G. D. McManus, J. M. Rawson, N. Feeder, J. Van Duijn, E. J. L. McInnes, J. J. Novoa, R. Burriel, F. Palacio and P. Oliete, *J. Mater. Chem.*, 2001, **11**, 1992–2003.

131   J. L. Brusso, O. P. Clements, R. C. Haddon, M. E. Itkis, A. A. Leitch, R. T. Oakley, R. W. Reed and J. F. Richardson, *J. Am. Chem. Soc.*, 2004, **126**, 8256–8265.

132   J. Zienkiewicz, P. Kaszynski and V. G. Young, *J. Org. Chem.*, 2004, **69**, 7525–7536.

133   J. Zienkiewicz, A. Fryszkowska, K. Zienkiewicz, F. Guo, P. Kaszynski, A. Januszko and D. Jones, *J. Org. Chem.*, 2007, **72**, 3510–3520.

Chapter 2

1,2,4,6-Thiatriazinyl Radicals and Dimers: Structural and Electronic Tuning Through Heteroaromatic Substituent Modification

2.1 Introduction

For some time stable radicals have been recognized as attractive building blocks in the development of molecular conductors and magnetic materials, in addition to their use as spin bearing ligands in transition metal complexes.[1–10] The exploration of organic and organo-main group radicals in the aforementioned applications represents a challenging yet fruitful research area, which drives the design and development of new open-shell materials. In that regard, heterocyclic thiazyl radicals are perhaps the most promising candidates as the S–N unit not only affords an unpaired electron, but the incorporation of heavy heteroatoms into planar π-conjugated frameworks also provides the potential for structural control through S···N′ and C–H···N′ interactions and better opportunities for enhanced magnetic and electronic interactions by virtue of greater orbital overlap.[11–13] These features have attracted researchers in the direction of incorporating S–N moieties into π-conjugated frameworks and a number of heterocyclic thiazyl radicals have been shown to crystallize as discrete radicals demonstrating interesting charge transport[14,15] and magnetic[15–18] properties. Others still have been reported to exhibit heat- and light-induced dimer-to-radical interconversions with potential applications in magnetic switches.[19,20]

Concerning their potential as spin bearing ligands, multidentate thiazyl radicals have proven to be an attractive building block.[21] In this regard, the thiatriazinyl (TTA) radical represents an archetypal system, as the general structure is ideal for the development of N-coordinate paramagnetic ligands. Furthermore, variation of the aromatic substituents, from phenyl groups to heterocycles containing nitrogen, oxygen or sulfur, facilitates the development of chelating paramagnetic ligands, thus enabling coordination to a potentially diverse range of transition metals and lanthanides. The first report of a TTA radical was described by Markovskii *et al.* using electron paramagnetic resonance (EPR) spectroscopy;[22] however, it was not until 1984 that Oakley and coworkers first isolated and structurally characterized 3,5-bisphenyl-1,2,4,6-thiatriazinyl (Ph$_2$TTA; Chart 2.1).[23] Since then, both symmetrically and asymmetrically substituted TTA radicals have been reported, most of which are partially functionalized by a phenyl group.[23–26] With respect to metal coordination, Boeré and coworkers have explored the reactivity of symmetric (Ph$_2$TTA) and asymmetric (3-phenyl-5-trifluoromethyl-TTA) derivatives with [Cr(Cp)(CO$_3$)$_2$].[27,28]As may be expected, the substituents played a role in the coordination as Ph$_2$TTA afforded an η^1–adduct bonded through the sulfur atom whereas the asymmetric ligand resulted in an η^2 S=N linkage. Interestingly, in both cases, coordination through the central nitrogen of the TTA ring did not occur. It is anticipated that heteroaromatic substituents, such as in 3,5-bis(2-pyridyl)-1,2,4,6-thiatriazinyl and 3,5-bis(2-thienyl)-1,2,4,6-thiatriazinyl (Py$_2$TTA and Th$_2$TTA, respectively; Chart 2.1), would promote such coordination through the chelating effect rendering the N–S–N portion of the heterocyclic ring devoid of substituents. This would therefore permit these heteroatoms to engage in supramolecular interactions, which has been shown to direct crystal packing in the solid state and facilitate conductive or magnetic exchange pathways in thiazyl based radical complexes.[21]

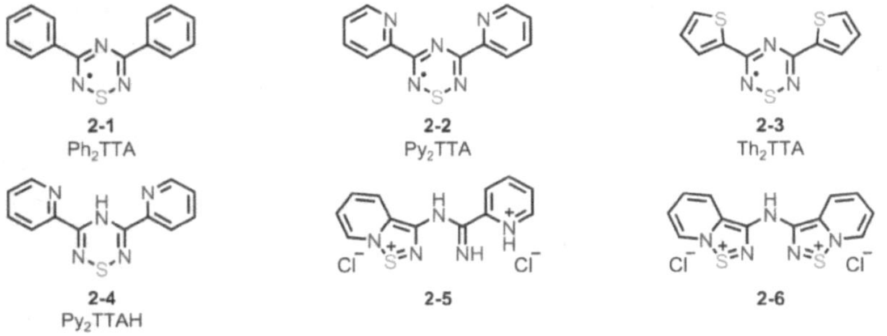

2-1
Ph$_2$TTA

2-2
Py$_2$TTA

2-3
Th$_2$TTA

2-4
Py$_2$TTAH

2-5

2-6

Chart 2.1 Examples of aryl-substituted TTA radicals and *N*-bridgehead thiadiazolium salts.

As a first step towards the development of heteroaromatic-functionalized TTA, the Brusso group reported the synthesis of 3,5-bis(2-pyridyl)-4*H*-1,2,4,6-thiatriazinyl (Py$_2$TTAH), a task which involved the intermediacy of the mono- and bis-*N*-bridgehead-1,2,5-thiadiazolium salts **2-5** and **2-6** (Chart 2.1).[29] This unusual non-innocent behavior of pyridyl nitrogen atoms has only rarely been observed;[30] thus the apparent proclivity of the pyridyl ligands to form *N*-bridgehead heterocycles represents an important finding and holds potential in the design of novel open- and closed-shell heterocyclic compounds. In that initial communication, the Py$_2$TTA radical was generated *in situ* and characterized by EPR spectroscopy.[29] We have now successfully isolated and characterized Py$_2$TTA in the solid state. In addition, the synthesis and characterization of the bisthienyl derivative Th$_2$TTA is also reported herein. Comparison of the EPR spectra of the two radicals, along with computational studies, reveals the impact of heteroaromatic substitution on the electronic structure of TTA radicals. Moreover, crystal analysis emphasizes the importance of intermolecular contacts on the crystal packing and demonstrates the potential for structural control through S$\cdots$N' and C–H$\cdots$N' interactions, which are significantly influenced by the presence of the pyridyl and thienyl groups in Py$_2$TTA and Th$_2$TTA, respectively. This work therefore represents a fundamental study of heterocyclic aromatic substitution in thiazyl-based radicals and investigates how varying these functional groups influences the molecular properties and long-range order within the supramolecular structure.

2.2 Results and Discussion

2.2.1 Synthesis and Isolation of Py₂TTA

Our initial synthetic work en route to the closed-shell Py₂TTAH elucidated the preparation and structural characterization of two unusual *N*-bridgehead-1,2,5-thiadiazolium salts **2-5** and **2-6**.[29] While the *in situ* preparation of Py₂TTA through oxidation of Py₂TTAH with half an equivalent of iodine in the presence of 4-dimethylaminopyridine (DMAP) was successfully achieved, isolation of the radical in the solid state proved difficult due to the formation of intractable mixtures of DMAP salts. Initial attempts to overcome this problem included deprotonation of Py₂TTAH prior to oxidation in order to isolate the anionic derivative; however, the proton affinity of the central nitrogen on the TTA ring is quite high and, regardless of the base used (e.g., DMAP, proton sponge, KOH, etc.), generation of the anion was unsuccessful. In fact, Py₂TTAH is surprisingly robust with a high tolerance towards aqueous, basic, and thermal conditions. Its only instability appears to be in acidic conditions, in which the thiatriazine ring opens and regenerates an *N*-bridgehead-1,2,5-thiadiazonium salt.[29]

Since deprotonation was not feasible, treatment with a relatively strong oxidant such as bromine was considered in order to prepare 3,5-bis(2-pyridyl)-1-bromo-1,2,4,6-thiatriazine. This was not only expected to eliminate the problem associated with the proton on the central TTA ring, but also to provide a compound that can be treated with a suitable reductant to afford Py₂TTA, similar to the aforementioned route employed for the phenyl-based derivatives.[23] However, while reaction with bromine formed the desired 1-bromo-1,2,4,6-thiatriazine, the generation of HBr led to protonation of the pyridyl nitrogen atoms affording the 3,5-bis(2-pyridinium)-1-bromo-1,2,4,6-thiatriazine dication, isolated as a mixed bromide/tribromide salt (**2-12**; Scheme 2.1). Crystallization of this material from dichloroethane (DCE) afforded yellow needles suitable for X-ray analysis confirming the structural identity of **2-12** (Figure 2.1a).

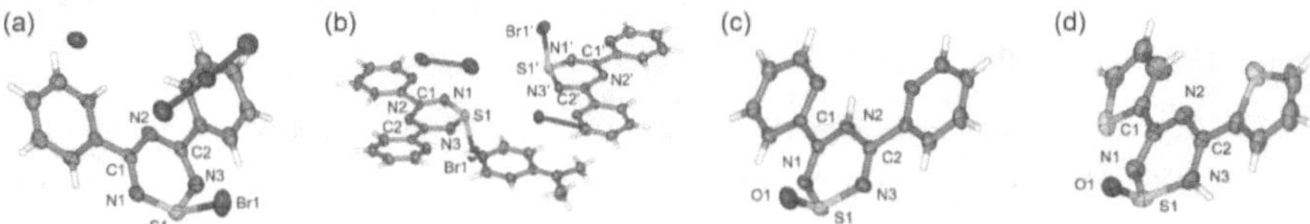

Figure 2.1 ORTEP drawings (50% thermal ellipsoids) of the asymmetric unit of (a) **2-12** (b) **2-13** (c) **2-14** and (d) **2-15** at 200 K, with atom numbering of the TTA ring (for crystallographic data see Tables A1.1 and A1.2 in Appendix A1). Disorder in the thiophenes of **2-15** is not shown.

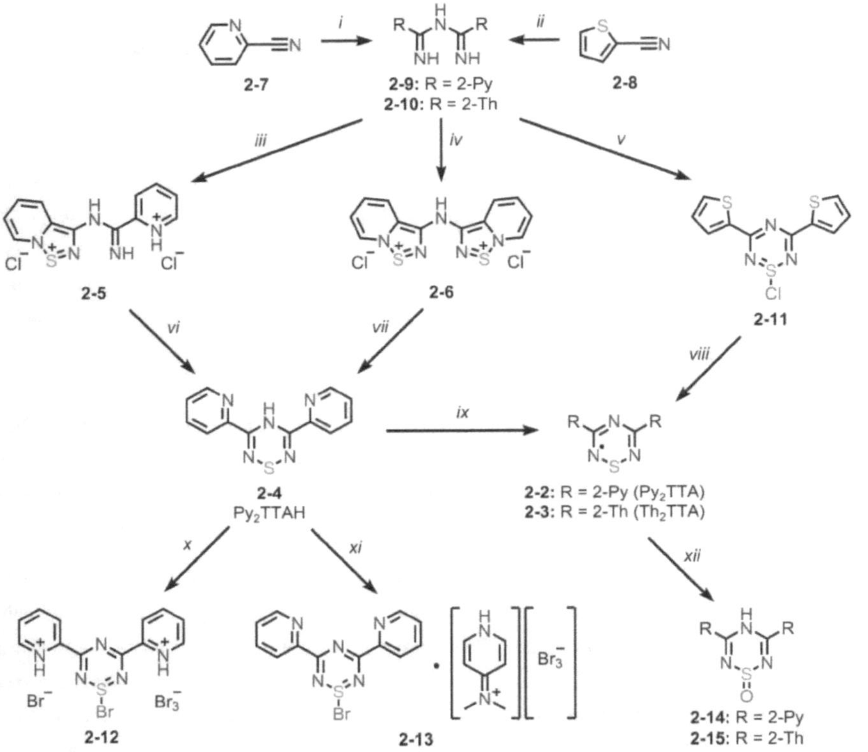

Scheme 2.1 Synthesis of Py$_2$TTA and Th$_2$TTA. Reagents and conditions: (*i*) NH$_{3(g)}$, MeCN, 110 °C, 72 h; (*ii*) 2-thiophenecarboxamidine, NaH, DMSO, 25 °C, 56 h; (*iii*) S$_2$Cl$_2$ (3 equiv.), MeCN, 25 °C, 3 h; (*iv*) S$_2$Cl$_2$ (4 equiv.), MeCN, reflux, 16 h; (*v*) S$_2$Cl$_2$ (5 equiv.), MeCN, reflux; (*vi*) PhCl, reflux, 72 h; (*vii*) SbPh$_3$, MeCN, reflux, 1 h; (*viii*) SbPh$_3$ (0.5 equiv.), MeCN, 25 °C, 1 h; (*ix*) NCS (0.5 equiv.), DMAP, MeCN, 25 °C, 2 h; (x) Br$_2$, MeCN, 25°C; (*xi*) Br$_2$, DMAP, MeCN, reflux; (*xii*) O$_2$, MeCN.

To avoid protonation of the pyridyl nitrogen atoms, Py$_2$TTAH was treated with bromine in the presence of DMAP, affording the desired compound 3,5-bis(2-pyridyl)-1-bromo-1,2,4,6-thiatriazine. However, while this confirmed that DMAP successfully inhibits protonation, the product **2-13** was isolated as a cocrystal with DMAP·HBr$_3$ (Figure 2.1b) and attempts to reduce this mixed salt with triphenylantimony resulted in the regeneration of Py$_2$TTAH. In order to isolate the Py$_2$TTA radical, we therefore turned to the use of oxidants, which would not generate protons as a side-product. Oxygen, of course, also had to be avoided due to the proclivity of TTA radicals to form 1-oxo-1,2,4,6-4*H*-thiatriazines (e.g., **2-14**; Figure 2.1c). Various mild oxidants, including 2,3-dichloro-5,6-dicyanobenzoquinone, *N*-bromosuccinimide and *N*-chlorosuccinimide (NCS) were explored, and of these NCS proved to be the most effective affording Py$_2$TTA as a microcrystalline copper colored solid. Recrystallization of Py$_2$TTA from hot acetonitrile (MeCN) afforded crystals suitable for X-ray analysis (*vide infra*).

2.2.2 Synthesis and Isolation of Th₂TTA

In regard to Th₂TTA, the absence of basic sites on the thienyl residue suggested that the synthesis would resemble that of the phenyl derivatives more closely than that of the pyridyl derivative. Nonetheless, regardless of the reaction pathway the most successful preparative routes include condensation of imidoylamidines with excess sulfur halides,[26,29,31] which therefore requires *N*-(2-thienylimidoyl)-2-thienylamidine (**2-10**) as a starting point. Although several preparative routes have been reported, unsubstituted *N*-imidoylamidines remain a synthetic challenge and have received relatively little attention in literature. The classical route involves the reaction between an *N*-thioimidoylamidine with an amidine,[32] whereas an alternative avenue includes ring opening of triazines with lithium bis(trimethylsilyl)amide.[33] More recently, however, nitriles activated through inductive effects have been shown to undergo nucleophilic addition reactions with ammonia or amidine.[26,29,34] Our initial attempts to prepare **2-10** therefore focused on the reaction of 2-cyanothiophene (**2-8**) with NH$_{3(g)}$, which proved successful for the pyridyl derivative (**2-9**).[29] Unfortunately, negligible yields of the desired product were obtained regardless of the temperature or pressure conditions used. This may be attributed to the electron donating character of the thiophene moiety, which deactivates the nitrile to nucleophilic addition. Similar challenges were encountered when effort was directed towards the reaction between the free-base thiophene-2-carboxamidine with 2-cyanothiophene. This was overcome by strengthening the nucleophilicity of the amidine through deprotonation *via* lithium diisopropylamide affording **2-10** as a brown oil. Although this route can be used to prepare **2-10**, isolation of *N*-(2-thienylimidoyl)-2-thienylamidine was problematic leading to relatively poor purified yields. This was alleviated by changing the base to sodium hydride and employing a very polar aprotic solvent (e.g., DMSO),[35] affording **2-10** in much higher yields and greater purity (Scheme 2.1). Subsequent condensation of **2-10** with excess sulfur monochloride afforded 3,5-bis(2-thienyl)-1-chloro-1,2,4,6-thiatriazine (**2-11**) as a bright yellow crystalline material. Due to its limited stability, **2-11** was used directly without further purification in the reduction reaction with triphenylantimony resulting in a dark green microcrystalline solid. Crystallization from hot DCE afforded Th₂TTA as dark red platelets suitable for single-crystal X-ray analysis; however, if molecular oxygen was present, crystals of **2-15** would be produced (Figure 2.1d).

2.2.3 EPR Spectroscopy of TTA Radicals

We previously reported the *in situ* EPR spectrum of Py$_2$TTA in dichloromethane (DCM), which consisted of a complex multiplet that was simulated using a model based on hyperfine coupling to two equivalent and one unique ^{14}N nuclei (experimentally derived constants: a_N = 3.72 G; a_N = 4.27 G; calculated coupling constants: a_N = 3.47 G; a_N = 4.20 G).[29] Interestingly, when recrystallized Py$_2$TTA was dissolved in degassed DCM, the solution remained EPR silent. While this finding was initially surprising, an EPR spectrum consistent with that obtained from *in situ* preparation of Py$_2$TTA was eventually achieved by switching to hot toluene (Figure 2.2a; a_N = 3.77 G and 4.38 G), which was necessary to provide the elevated temperatures required to dissolve Py$_2$TTA. By contrast, the X-band EPR spectrum of Th$_2$TTA recorded in DCM at room temperature was much more easily obtained, indicating that dissolution and/or dissociation of Th$_2$TTA is clearly more extensive than in the case of Py$_2$TTA. The EPR for Th$_2$TTA exhibits a seven-line pattern that is consistent with the DFT calculated spin distribution for a TTA radical (Figure 2.2b). It was simulated using a model based on hyperfine coupling to three quasi-equivalent ^{14}N nuclei (a_N = 3.92 G and 3.86 G), which may be compared with DFT-calculated values (a_N = 3.77 G and 3.67 G). The near equivalence of the a_N values here is in contrast to Py$_2$TTA (Figure 2.2a) and is consistent with the expected electronic differences between the thienyl and pyridyl ligands, the latter being a more potent electron-withdrawing group. As a result, the Py$_2$TTA spin density is more extensively polarized away from the N–S–N region of the TTA core. Similar trends have been observed in the EPR spectra or *p*-substituted aryl TTA radicals.[25,36] For example, spin density in TTA systems is equally distributed over the three nitrogen atoms when phenyl or 4-MeOC$_6$H$_4$ substituents are employed, but becomes strongly polarized toward the central nitrogen atom (in the TTA ring) when electron-withdrawing 4-NO$_2$C$_6$H$_4$ groups are present.[25] Thus comparison between the EPR spectra of Py$_2$TTA and Th$_2$TTA confirms the expectation that heteroaromatic substitution influences the electronic structure of TTA radicals. Furthermore, this study also speaks to the solubility and, accordingly, the strength of intermolecular interactions observed in the pyridyl derivative compared to that of Th$_2$TTA, which is further exemplified by inspection of the crystal structures (*vide infra*).

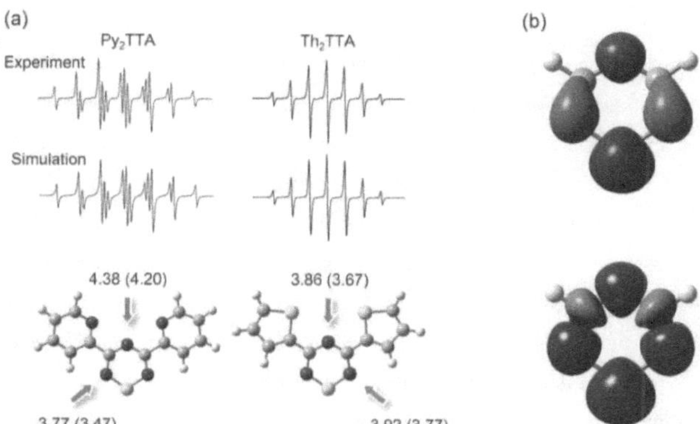

Figure 2.2 (a) Experimental (top) and simulated (middle) EPR spectra of (left) Py$_2$TTA in toluene collected at 40 °C (g = 2.0068; SW = 35 G; LW = 0.24 G) and (right) Th$_2$TTA in DCM collected at RT (g = 2.0066; SW = 35 G; LW = 0.19 G). Experimentally derived and calculated (in parentheses) coupling constants a_N in G (below). (b) UB3LYP/6-311+G(2d,p) calculated (top) SOMO and (bottom) spin density distribution for a geometry optimized model of 1,2,4,6-thiatriazinyl.

2.2.4 Structural Analysis of TTA Radicals

To explore the impact of heteroaromatic substitution (i.e., pyridyl *vs.* thienyl) on crystal packing, single crystals of Py$_2$TTA and Th$_2$TTA were analyzed by X-ray diffraction. Crystallographic data and selected bond lengths are listed in Table A1.3 and A1.4 of Appendix A1, while ORTEP drawings of the molecules in the asymmetric units are provided in Figure 2.3.

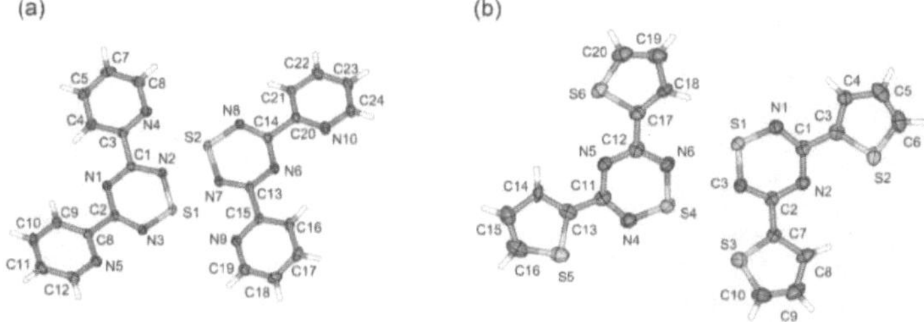

Figure 2.3 ORTEP drawings (50% thermal ellipsoids) of the asymmetric unit of (a) Py$_2$TTA and (b) Th$_2$TTA at 200 K, with atom numbering. Disorder in the thiophenes of Th$_2$TTA is not shown. Colour code: C, grey; H, white; N, blue; S, yellow.

Two views of the solid-state structures, that of the cofacial π-dimer and dimer-of-dimers (i.e., tetramers), are shown in Figure 2.4. Focusing on the molecular framework, a number of similarities exist between the two compounds. For example, crystals of Py$_2$TTA and Th$_2$TTA consist of discrete pairs of TTA rings linked in a cofacial manner containing short S$\cdots$S contacts of 2.592(6) Å and 2.647(1) Å, respectively. While these distances are longer than a disulfide bond (2.06 Å in S$_8$),[37] they are substantially shorter than the sum of the van der Waals radii (3.6 Å)[38] and are comparable to the transannular S$\cdots$S' interactions often observed in binary sulfur nitrides, which range from 2.43 to 2.65 Å (e.g., 2.43 Å for 3,7-bis(dimethylamino)-1,5,2,4,6,8-dithiatetrazocine;[39] 2.58 Å for S$_4$N$_4$).[40] As well, since the two TTA rings are perfectly eclipsed, numerous close S$\cdots$N', N$\cdots$N' and C$\cdots$C' interactions exist within the dimer as a result of the close S$\cdots$S' contact. In the structure of both Py$_2$TTA and Th$_2$TTA, the two TTA rings that form the cofacial dimer exhibit shallow boat-like configurations due to a slight displacement of the nitrogen and sulfur atoms that reside along the center of the molecular framework (i.e., the sulfur and nitrogen atom para to it). It is interesting to note that similar structural features (e.g., cofacial dimer, boat-like configuration) were described for the phenyl derivative Ph$_2$TTA;[23] indicative of minimal structural variation of the TTA ring and cofacial dimers upon heterocyclic substitution.

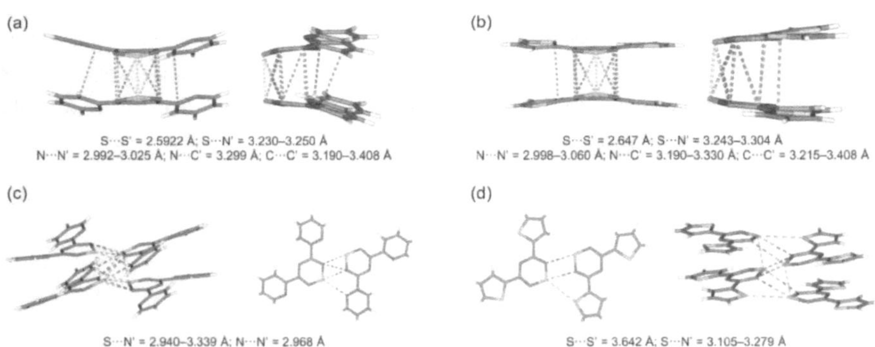

(a)

S$\cdots$S' = 2.5922 Å; S$\cdots$N' = 3.230–3.250 Å
N$\cdots$N' = 2.992–3.025 Å; N$\cdots$C' = 3.299 Å; C$\cdots$C' = 3.190–3.408 Å

(b)

S$\cdots$S' = 2.647 Å; S$\cdots$N' = 3.243–3.304 Å
N$\cdots$N' = 2.998–3.060 Å; N$\cdots$C' = 3.190–3.330 Å; C$\cdots$C' = 3.215–3.408 Å

(c)

S$\cdots$N' = 2.940–3.339 Å; N$\cdots$N' = 2.968 Å

(d)

S$\cdots$S' = 3.642 Å; S$\cdots$N' = 3.105–3.279 Å

Figure 2.4 Two views of the dimer (a, b) and dimer-of-dimers (c, d) for Py$_2$TTA (a, c) and Th$_2$TTA (b, d) are shown with close intermolecular contacts highlighted by dashed lines. Faded regions indicate molecules in background of view. Colour code: C, grey; H, white; N, blue; S, yellow.

While a number of similarities exist between the structures of Py$_2$TTA and Th$_2$TTA, the impact of heteroaromatic substitution becomes apparent when looking at the supramolecular contacts, which direct packing in the solid state. For example, within the asymmetric unit of Py$_2$TTA, which belongs to the triclinic space group $P\bar{1}$ and contains two molecules, many unique close contacts that are within or nominally larger than the sum of the van der Waals separation exist,[38,41] whereas crystals of Th$_2$TTA belong to the monoclinic space group $P2_1/n$ and consists of fewer intermolecular interactions. Although both compounds form tetramers in the solid state, in Py$_2$TTA these head-to-head dimer-of-dimers are held together by numerous S$\cdots$N' interactions, whereas only one S$\cdots$S' and three unique S$\cdots$N' contacts exist in the structure of Th$_2$TTA. Furthermore, the tetramers of Py$_2$TTA form a brick-like array held together by a series of close π-contacts between the carbon atoms on the pyridyl rings and either the pyridyl or TTA ring on adjacent tetramers (Figure 2.5a). A bird's-eye view of the Py$_2$TTA brick-like array reveals additional close contacts, many of which stem from hydrogen bonding between adjacent pyridyl rings, holding the molecules together in a 2D fashion (Figure 2.5b).

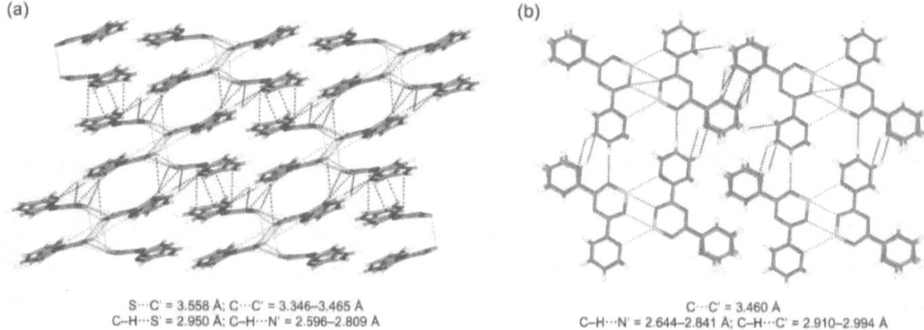

Figure 2.5 Views of the packing arrangement exhibited by crystals of Py$_2$TTA demonstrating (a) the brick-like array of π-stacked tetramers held by interlayer π–π contacts and (b) the two-dimensional network formed between tetramers of adjacent brick-like arrays linked by lateral contacts. Contacts lacing dimer-of-dimers (including S$\cdots$S' dimer contacts) are shown in yellow dashed lines and contacts between tetramers are shown in red dashed lines. Colour code: C, grey; H, white; N, blue; S, yellow.

Similar to Py$_2$TTA, tetramers of Th$_2$TTA also form a brick-like arrangement held together primarily through π–π interactions (e.g., C$\cdots$C' and C$\cdots$S'); however, contrary to Py$_2$TTA, the structure of Th$_2$TTA consists of alternating layers of π-stacked tetramers that are rotated with respect to one another by 51.60(7)° forming a cross-braced configuration (Figure 2.6a). These layers are held together through lateral C–H$\cdots$N', C–H$\cdots$S, and S$\cdots$C' interactions (Figure 2.6b). While a number of contacts exist within the thienyl derivative, there are fewer than that of the pyridyl derivative. Such structural features may explain the enhanced solubility and presence of an EPR signal for Th$_2$TTA compared with Py$_2$TTA, which is EPR silent at room temperature. While the electron-withdrawing vs. donating substituents have little structural impact on the molecular framework of the thiatriazinyl rings, heteroaromatic substitution (i.e., pyridyl vs. thienyl) significantly influences the long-range molecular order and, therefore, the supramolecular

41

architectures. This finding is further supported by comparing the crystal packing of the phenyl derivative (Ph₂TTA) with that of the pyridyl (Py₂TTA) and thienyl (Th₂TTA) analogs, where even fewer supramolecular contacts exist.[23]

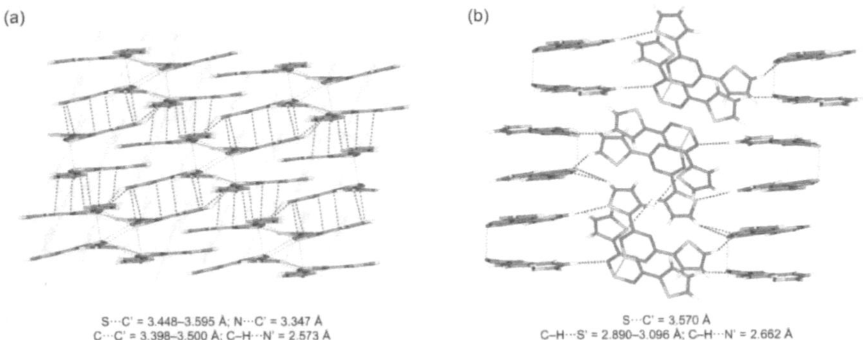

Figure 2.6 (a) View of the packing arrangement exhibited by crystals of Th₂TTA demonstrating (a) the brick-like arrays of π-stacked tetramers held by interlayer π–π contacts and the cross-braced arrangement between adjacent arrays. Faded regions indicate molecules in background of view. (b) Orthogonal viewpoint of adjacent arrays linked by lateral contacts. Contacts lacing dimer-of-dimers (including S⋯S' dimer contacts) are shown in yellow dashed lines and contacts between tetramers are shown in red dashed lines. Colour code: C, grey; H, white; N, blue; S, yellow.

2.3 Summary and Conclusion

The isolation and solid-state characterization of Py₂TTA has now been achieved. In addition, the thienyl derivative Th₂TTA was prepared and structurally characterized. From a synthetic standpoint, the nature of the heteroatom in the aromatic substituents plays a pivotal role in the preparative pathway. Based on the results described here, the generation of a wide range of heteroaromatic TTA derivatives can be anticipated. Comparison of the EPR spectra along with computational studies demonstrates the effect of heteroaromatic substitution on the electronic structure, while crystal analysis emphasizes the importance of intermolecular contacts on the crystal packing. In this regard, although structural studies reveal several similarities between Py₂TTA and Th₂TTA, the presence of heteroatoms in the aromatic substituents significantly influences the crystal packing and, hence, the supramolecular architectures. The enhanced intermolecular interactions observed in Py₂TTA and Th₂TTA (compared to Ph₂TTA) result in more tightly bound structures, the consequence of which was particularly noted in the EPR studies of Py₂TTA, in which a purified sample requires hot toluene to produce an EPR signal.

Inclusion of heteroatoms in the aromatic substituents of TTA not only influences the molecular and solid-state properties of the discrete molecules, but it also facilitates coordination to a variety of metals. Not only is this anticipated to stabilize the unpaired electron thus preventing dimerization, but it also provides an avenue to generate coordination complexes that can take advantage of the unencumbered N–S–N portion of the TTA ring, which can engage in supramolecular interactions that are known to direct crystal packing in thiazyl based ligands and provide conductive and magnetic exchange pathways. In addition, the redox-flexible properties of TTA enables a degree of tunability in regard to the oxidation state of the ligand framework (e.g., 7π-neutral radical and 8π-anion) prior to or post-coordination. Thus, the synthetic preparations described here presents new opportunities to explore heteroaromatic-substituted TTAs for the development of organic-metal hybrid materials.

2.4 Experimental Details

2.4.1 General Procedures

The reagents sulfur monochloride (Sigma-Aldrich) and triphenylantimony (TCI America) were obtained commercially and used as received. Sodium hydride (60% dispersion in mineral oil, Alfa Aesar) was washed with hexanes prior to use. **2-4**,[29] **2-8**,[42] and 2-thiophenecarboxamidine[43] were prepared as outlined in the literature. All solvents were of reagent grade; MeCN and DCE were dried by distillation over P_2O_5 and stored over 4 Å molecular sieves. Dimethyl sulfoxide was dried and stored over 4 Å molecular sieves. Unless otherwise indicated, all reactions were performed under an atmosphere of dry nitrogen using dried solvents. Melting points were taken using a Mel-Temp apparatus and are uncorrected. NMR (^{1}H, ^{13}C) spectroscopy was performed on samples dissolved in deuterated solvents at room temperature using a Bruker Avance 400 MHz spectrometer and the spectra were referenced to the residual solvent signal.[44] FTIR spectroscopy was performed on solid samples using an Agilent Technologies Cary 630 FTIR spectrometer. Elemental analyses were performed by MHW Laboratories, Phoenix, AZ 85018.

2.4.2 Experimental Procedures

Preparation of 3,5-bis(2-pyridyl)-1,2,4,6-thiatriazinyl (2-2). Py$_2$TTAH (0.500 g, 1.96 mmol), 4-dimethylaminopyridine (0.136 g, 1.11 mmol) and N-chlorosuccinimide (0.134 g, 1.00 mmol) were combined with degassed MeCN (5 mL; 3 freeze-pump-thaw cycles) affording a maroon slurry. After 2 h, a fine brown solid was filtered *in vacuo* and washed twice with MeCN (3 mL). Crude yield = 0.175 g (0.688 mmol, 35%). Recrystallization from MeCN afforded black-brown microcrystalline material of **2-2**. IR (υ_{max}, ATR): 3050 (w), 3004 (w), 1641 (w), 1571 (w), 1530 (m), 1510 (m), 1489 (m), 1476 (m), 1461 (m), 1422 (m), 1396 (m), 1380 (s), 1345 (m), 1254 (m), 1231 (m), 1160 (w), 1112 (m), 1042 (w), 994 (m), 974 (w), 912 (w), 843 (w), 814 (w), 768 (s), 757 (m), 739 (s), 726 (s), 692 (m), 668 (w) cm^{-1}. Anal. Calcd for $C_{12}H_8N_5S$: C, 56.68; H, 3.17; N, 27.54%. Found: C, 56.90; H, 3.37; N, 27.66%.

Preparation of *N*-(2-thienylimidoyl)-2-thienylamidine (2-10). Sodium hydride (1.71 g, 71.3 mmol) was added to a solution of 2-thiophenecarboxamidine (6.00 g, 47.6 mmol) and 2-thiophenecarbonitrile (5.20 g, 47.6 mmol) in DMSO (86 mL). After stirring for 56 h at room temperature, the resulting dark brown solution was quenched with 200 mL of water and extracted with DCM. The organic phase was washed with water, dried over potassium carbonate and filtered. The filtrate was dried *in vacuo* affording **2-10** as a brown solid (9.26 g, 39.4 mmol, 83%). Pale yellow needles of **2-10** were obtained by recrystallization in hexanes. MP = 98–99 °C. ^{1}H NMR (δ, CDCl$_3$): 7.51 (2H, dd, J = 3.8, 1.2 Hz), 7.46 (2H, dd, J = 5.0, 1.2 Hz), 7.09 (2H, dd, J = 5.0, 3.8 Hz). ^{13}C NMR (δ, CDCl$_3$): 162.09 (2C), 142.75 (2C), 130.10 (2CH), 127.73 (2CH), 126.43 (2CH). IR (υ_{max}, ATR): 3435 (m), 3263 (m), 1588 (s), 1523 (s), 1481 (s), 1423 (s), 1382 (s), 1320 (m), 1241 (w), 1167 (s), 1116 (w), 1047 (m), 978 (m), 910 (w), 856 (m), 827 (m) cm^{-1}. Anal. calcd for C$_{10}$H$_9$N$_3$S$_2$: C, 51.04; H, 3.86; N, 17.86%. Found: C, 51.16; H, 3.73; N, 17.74%.

Preparation of 1-chloro-3,5-bis(2-thienyl)-1λ^4,2,4,6-thiatriazine (2-11). Sulfur monochloride (5.72 g, 42.4) was added dropwise to a stirring solution of **2-10** (2.0 g, 8.50 mmol) in MeCN (50 mL) forming a white precipitate, which dissolved upon heating to reflux. The resulting bright orange solution was filtered hot and, on cooling to 0 °C, yellow needle-like crystals formed. The crystals were filtered and dried *in vacuo*. Yield 2.07 g (7.22 mmol, 85%). ^{1}H NMR (δ, CDCl$_3$): 8.22 (2H, dd, J = 3.9, 1.3 Hz), 7.78 (2H, dd, J = 5.0, 1.3 Hz), 7.24 (2H, dd, J = 5.0, 3.9 Hz), ^{13}C NMR (δ, CDCl$_3$): 165.48 (2C), 140.85 (2C), 135.91 (2CH), 134.38 (2CH), 129.09 (2CH). IR (υ_{max}, ATR): 3102 (w), 1516 (s), 1478 (w), 1438 (m), 1396 (s), 1219 (s), 1113 (m), 1027 (br, s), 944 (br, m), 858 (s), 838 (m), 823 (m), 768 (s), 745 (m), 715 (s), 699 (s) cm^{-1}.

Preparation of 3,5-bis(2-thienyl)-1,2,4,6-thiatriazinyl (2-3). Triphenylantimony (1.22 g, 3.45 mmol) and **2-11** (2.07 g, 6.90 mmol) were combined with degassed MeCN (3 freeze-pump-thaw cycles) resulting in a maroon slurry. After stirring for 1 h, a dark purple solid was filtered in vacuo. Crude yield 1.66 g (3.14 mmol, 91%). Recrystallization from DCE afforded **2-3** as sparkly, green plates. dec >135 °C. IR (υ_{max}, ATR): 3089 (w), 1519 (m), 1464 (m), 1418 (s), 1404 (s), 1370 (s), 1340 (s), 1328 (s), 1236 (m), 1196 (s), 1142 (w), 1126 (w), 1080 (m), 1052 (m), 1031 (s), 1015 (s), 1001 (s), 940 (m), 910 (w), 863 (m), 853 (m), 798 (m), 764 (s), 747 (s), 737 (s), 710 (s) cm^{-1}. Anal. calcd for C$_{10}$H$_6$N$_3$S$_3$: C, 45.43; H, 2.29; N, 15,90%. Found: C, 45.60; H, 2.47; N, 15.97%.

2.4.3 EPR Spectroscopy

The X-band EPR spectra of Py$_2$TTA and Th$_2$TTA were recorded using a Bruker EMX-200 spectrometer on samples dissolved in degassed dichloromethane and/or toluene. Hyperfine coupling constants were obtained by spectral simulation using PEST Winsim and Bruker WinEPR Simfonia.[44]

2.4.4 Crystallography

The data collection results for compounds **2-12**, **2-13**, **2-14**, **2-15**, Py$_2$TTA and Th$_2$TTA represent the best data sets obtained in several trials for each sample. The crystals were mounted on thin glass fibers using paraffin oil. Prior to data collection, crystals were cooled to 200.15 K. The data was collected on a Bruker AXS KAPPA single crystal diffractometer equipped with a sealed Mo tube source (wavelength 0.71073 Å) and APEX II CCD detector. The raw data collection and processing were performed with the APEX II software package from BRUKER AXS.[45] The diffraction data for crystals of **2-12**, **2-15** and Th$_2$TTA were collected with a sequence of 0.3° ω scans at 0, 120, and 240° in φ. Due to the lower unit cell symmetry, in order to ensure adequate data redundancy, diffraction data for **2-13**, **2-14** and Py$_2$TTA were collected with a sequence of 0.3° ω scans at 0, 90, 180 and 270° in φ. The initial unit cell parameters were determined from 60 data frames with 0.3° ω scan each, collected at the different sections of the Ewald sphere. Semi-empirical absorption corrections based on equivalent reflections were applied.[46] Systematic absences in the diffraction data and unit cell parameters were consistent with triclinic $P\bar{1}$ for compound **2-13**, **2-14** and Py$_2$TTA, monoclinic $P2_1/c$ for compound **2-12**, monoclinic $P2_1$ for **2-15** and monoclinic $P2_1/n$ for Th$_2$TTA. Solutions in the centrosymmetric space groups for all the compounds except **2-15** yielded chemically reasonable and computationally stable results of refinement. Meanwhile, diffraction data for **2-15** suggested non-centrosymmetric space groups for the structural model. The structures were solved by direct methods, completed with difference Fourier synthesis, and refined with full-matrix least-squares procedures based on F^2.

The structural model for **2-14** contains one target compound molecule located in general positions. An initial unit cell determination procedure suggested that the crystal mounted on the machine was non-merohedrally twinned. Refinement of the structure confirmed the presence of the second independent crystallographic domain. In order to find independent orientation matrices 5423 reflections were collected from 3 sets of 50 frames each in the different sections of the Ewald sphere. The collected reflection data were processed with CELL_NOW software[45] and produced two independent orientation matrices. The data set was re-integrated with two obtained matrices, treated for twinning absorption corrections and consecutive model refinement was performed against HKLF 5 reflection data file. Twinning domain ratio coefficient (BASF) was refined to 0.4102.

The asymmetric unit for compound **2-13** includes two target molecules and one molecule of *para*-amino pyridine, all located in general positions. The structure also incorporates two Br$_3^-$ anions located in the two independent inversion centers.

The diffraction data for the crystal of the complex **2-12** were collected to 0.75Å resolution; The absorption correction stage, that both R(int) and R(sigma) exceed 35% for the data below 0.85Å resolution. Based on the R(sigma) value, data were truncated to 0.80Å resolution for refinement. The asymmetric unit for this crystallographic model of **2-12** consists of one target compound molecule, one Br_3^- anion and one separate Br^- atom. All the fragments are located in the general positions.

The structural model of Py_2TTA consists of two aligned molecules of the target compound with close contact between two sulfur atoms of the units (S(1)–S(2) = 2.59Å). Both molecules are located in the general positions in the unit cell.

The asymmetric unit for the structure of **2-15** contains one target molecule located in a general position. On the final stages of refinement, the numbers suggested the presence of merohedral twinning in the structure. To take this into account, simple TWIN and BASF instructions were introduced and the twinning parameter was refined to 0.0687. After introduction of anisotropic refinement parameters for all the non-hydrogen atoms, it was discovered that one of the 5-member rings (S(3), C(7)–C(9)) of the structure is disordered by 180°-rotation around the C(2)–C(7) carbon-carbon bond. The atomic positions of this fragment were split to model the disorder and occupancy was allowed to refine. On the latest stages of refinement, occupancy was fixed at 61%–39% providing satisfactory anisotropic thermal motion parameters. To ensure proper geometry of the disordered moiety, a set of geometric restrains (SAME) were introduced.

The structural model for Th_2TTA demonstrates two molecules of the target compound located in the general position of the unit cell. Similar to the structure of Py_2TTA, these two molecules reveal close contact between two sulfur atoms (S(1)–S(4) = 2.65Å). The anisotropic refinement of the structure revealed possible disorder by 180° rotation around the carbon bonds for all four of the 5-member rings. To model the disorder, the positions of sulfur atoms within these 5-member rings and the positions of carbon atoms opposite to sulfur were split and the occupancy coefficients were allowed to refine. On the later stages of refinement, occupancy coefficients were fixed at the following values: S(2), C(3)-C(6) ring is rotated around C(1)–C(3) bond with partial occupancies 90% : 10%; S(3), C(7)–C(10) ring is rotated around C(2)–C(7) bond with partial occupancies 75% : 25%; S(5), C(13)–C(16) ring is rotated around C(11)–C(13) bond with partial occupancies 80% : 20%; S(6), C(17)–C(20) ring is rotated around C(12)–C(17) bond with partial occupancies 65% : 35%. To ensure proper fragments, geometries and acceptable anisotropic thermal displacement coefficients for two sets of geometry (SADI) and thermal motions (EADP) restraints were used during the refinement.

For all six compounds, positions of all hydrogen atoms were calculated based on the geometry of related non-hydrogen atoms. All hydrogen atoms were treated as idealized contributions during the refinement. All scattering factors are contained in several versions of the SHELXTL program library, with the latest version used being v.6.12.[47]

2.4.5 Theoretical Calculations

All calculations were performed at the DFT level using the UB3LYP functional, as contained in the Gaussian 09W suite of programs.[48] Molecular geometries were optimized using the 6-311+G(2d,p) basis set without symmetry constraints. Isotropic Fermi contacts were taken from single-point calculations using the double-zeta basis set EPR-II for first-row atoms and split-valence triple-basis set 6-311G(d,p) for second-row atoms on the optimized geometries.

2.5 References

1   R. G. Hicks, in *Stable Radicals: Fundamentals and Applied Aspects of Odd-Electron Compounds*, John Wiley & Sons, Ltd, Chichester, UK, 2010, pp. 317–380.

2   I. Ratera and J. Veciana, *Chem. Soc. Rev.*, 2012, **41**, 303–349.

3   R. G. Hicks, *Nat. Chem.*, 2011, **3**, 189–191.

4   A. W. Cordes, R. C. Haddon and R. T. Oakley, *Phosphorus. Sulfur. Silicon Relat. Elem.*, 2004, **179**, 673–684.

5   J. M. Rawson, A. Alberola and A. Whalley, *J. Mater. Chem.*, 2006, **16**, 2560.

6   A. Jankowiak, D. Pociecha, J. Szczytko, H. Monobe and P. Kaszyński, *J. Am. Chem. Soc.*, 2012, **134**, 2465–2468.

7   J. W. L. Wong, A. Mailman, K. Lekin, S. M. Winter, W. Yong, J. Zhao, S. V. Garimella, J. S. Tse, R. A. Secco, S. Desgreniers, Y. Ohishi, F. Borondics and R. T. Oakley, *J. Am. Chem. Soc.*, 2014, **136**, 1070–1081.

8   K. Awaga, K. Nomura, H. Kishida, W. Fujita, H. Yoshikawa, M. M. Matsushita, L. Hu, Y. Shuku and R. Suizu, *Bull. Chem. Soc. Jpn.*, 2014, **87**, 234–249.

9   K. E. Preuss, *J. Chem. Soc. Dalt. Trans.*, 2007, 2357–2369.

10   M. Sorai, Y. Nakazawa, M. Nakano and Y. Miyazaki, *Chem. Rev.*, 2013, **113**, PR41–PR122.

11   S. Steinberger, A. Mishra, E. Reinold, J. Levichkov, C. Uhrich, M. Pfeiffer and P. Bäuerle, *Chem. Commun.*, 2011, **47**, 1982–1984.

12   T. Kono, D. Kumaki, J. I. Nishida, S. Tokito and Y. Yamashita, *Chem. Commun.*, 2010, **46**, 3265–3267.

13   M. Karikomi, C. Kitamura, S. Tanaka and Y. Yamashita, *J. Am. Chem. Soc.*, 1995, **117**, 6791–6792.

14   A. Mailman, S. M. Winter, X. Yu, C. M. Robertson, W. Yong, J. S. Tse, R. A. Secco, Z. Liu, P. A. Dube, J. A. K. Howard and R. T. Oakley, *J. Am. Chem. Soc.*, 2012, **134**, 9886–9889.

15   A. A. Leitch, K. Lekin, S. M. Winter, L. E. Downie, H. Tsuruda, J. S. Tse, M. Mito, S. Desgreniers, P. A. Dube, S. Zhang, Q. Liu, C. Jin, Y. Ohishi and R. T. Oakley, *J. Am. Chem. Soc.*, 2011, **133**, 6051–6060.

16   S. M. Winter, A. Mailman, R. T. Oakley, K. Thirunavukkuarasu, S. Hill, D. E. Graf, S. W. Tozer, J. S. Tse, M. Mito and H. Yamaguchi, *Phys. Rev. B*, 2014, **89**, 214403.

17   S. M. Winter, S. Datta, S. Hill and R. T. Oakley, *J. Am. Chem. Soc.*, 2011, **133**, 8126–8129.

18   A. Alberola, R. J. Less, C. M. Pask, J. M. Rawson, F. Palacio, P. Oliete, C. Paulsen, A. Yamaguchi, R. D. Farley and D. M. Murphy, *Angew. Chemie Int. Ed.*, 2003, **42**, 4782–4785.

19   K. Lekin, H. Phan, S. M. Winter, J. W. L. Wong, A. A. Leitch, D. Laniel, W. Yong, R. A. Secco, J. S. Tse, S. Desgreniers, P. A. Dube, M. Shatruk and R. T. Oakley, *J. Am. Chem. Soc.*, 2014, **136**, 8050–8062.

20   H. Phan, K. Lekin, S. M. Winter, R. T. Oakley and M. Shatruk, *J. Am. Chem. Soc.*, 2013, **135**, 15674–15677.

21   K. E. Preuss, *Coord. Chem. Rev.*, 2015, **289–290**, 49–61.

22   L. N. Markovskii, P. P. Kornuta, L. S. Kachkovskaya and O. M. Polumbrik, *Sulfur Lett.*, 1983, **1**, 143–145.

23   P. J. Hayes, R. T. Oakley, A. W. Cordes and W. T. Pennington, *J. Am. Chem. Soc.*, 1985, **107**, 1346–1351.

24   A. W. Cordes, P. J. Hayes, P. D. Josephy, H. Koenig, R. T. Oakley and W. T. Pennington, *J. Chem. Soc., Chem. Commun.*, 1984, 1021–1022.

25   R. T. Boere, R. T. Oakley, R. W. Reed and N. P. C. Westwood, *J. Am. Chem. Soc.*, 1989, **111**, 1180–1185.

26   R. T. Boeré, T. L. Roemmele and X. Yu, *Inorg. Chem.*, 2011, **50**, 5123–5136.

27   C. Y. Ang, R. T. Boeré, L. Y. Goh, L. L. Koh, S. L. Kuan, G. K. Tan and X. Yu, *Chem. Commun.*, 2006, 4735–4737.

28   C. Y. Ang, S. L. Kuan, G. K. Tan, L. Y. Goh, T. L. Roemmele, X. Yu and R. T. Boeré, *Can. J. Chem.*, 2015, **93**, 181–195.

29   A. A. Leitch, I. Korobkov, A. Assoud and J. L. Brusso, *Chem. Commun.*, 2014, **50**, 4934–4936.

30   C. E. Bacon, D. J. Eisler, R. L. Melen and J. M. Rawson, *Chem. Commun.*, 2008, 4924–4926.

31   R. T. Oakley, R. W. Reed, A. W. Cordes, S. L. Craig and J. B. Graham, *J. Am. Chem. Soc.*, 1987, **109**, 7745–7749.

32   D. A. Peak, *J. Chem. Soc.*, 1952, 215–226.

33   A. W. Cordes, C. D. Bryan, S. R. Scott, W. M. Davis, R. H. de Laat, J. D. Goddard, R. G. Hicks, D. K. Kennepohl, R. T. Oakley, N. P. C. Westwood, S. H. Glarum and R. C. Haddon, *J. Am. Chem. Soc.*, 1993, **115**, 7232–7239.

34   H. C. Brown and P. D. Schuman, *J. Org. Chem.*, 1963, **28**, 1122–1127.

35   J. Bisson, J. Dehaudt, M.-C. Charbonnel, D. Guillaneux, M. Miguirditchian, C. Marie, N. Boubals, G. Dutech, M. Pipelier, V. Blot and D. Dubreuil, *Chem. - A Eur. J.*, 2014, **20**, 7819–7829.

36   R. T. Boeré and T. L. Roemmele, *Phosphorus. Sulfur. Silicon Relat. Elem.*, 2004, **179**, 875–882.

37   B. Meyer, *Chem. Rev.*, 1976, **76**, 367–388.

38   Y. V. Zefirov and P. M. Zorkii, *J. Struct. Chem.*, 1976, **17**, 851–854.

39   I. Ernest, W. Holick, G. Rihs, D. Schomburg, G. Shoham, D. Wenkert and R. B. Woodwardt, *J. Am. Chem. Soc.*, 1981, **103**, 1540–1544.

40   B. D. Sharma and J. Donohue, *Acta Crystallogr.*, 1963, **16**, 891–897.

41   Y. V. Zefirov and P. M. Zorkii, *J. Struct. Chem.*, 1977, **17**, 644–645.

42   A. K. Chakraborti, G. Kaur and S. Roy, *Indian J. Chem. - Sect. B Org. Med. Chem.*, 2001, **40**, 1000–1006.

43   N. Zhang, S. Ayral-Kaloustian, T. Nguyen, R. Hernandez, J. Lucas, C. Discafani and C. Beyer, *Bioorganic Med. Chem.*, 2009, **17**, 111–118.

44   G. R. Fulmer, A. J. M. Miller, N. H. Sherden, H. E. Gottlieb, A. Nudelman, B. M. Stoltz, J. E.

Bercaw and K. I. Goldberg, *Organometallics*, 2010, **29**, 2176–2179.

45   I. Bruker Instruments, 1996.

46   Bruker AXS Inc., *APEX II*, 2007.

47   R. H. Blessing, *Acta Crystallogr. Sect. A*, 1995, **51**, 33–38.

48   G. M. Sheldrick, *Acta Crystallogr. Sect. A Found. Crystallogr.*, 2008, 64, 112–122.

49   M. J. Frisch, G. W. Trucks, H. B. Schlegel, G. E. Scuseria, M. A. Robb, J. R. Cheeseman, G. Scalmani, V. Barone, B. Mennucci, G. A. Petersson, H. Nakatsuji, M. Caricato, X. Li, H. P. Hratchian, A. F. Izmaylov, J. Bloino, G. Zheng, J. L. Sonnenberg, M. Hada, M. Ehara, K. Toyota, R. Fukuda, J. Hasegawa, M. Ishida, T. Nakajima, Y. Honda, O. Kitao, H. Nakai, T. Vreven, J. Montgomery, J. A., J. E. Peralta, F. Ogliaro, M. Bearpark, J. J. Heyd, E. Brothers, K. N. Kudin, V. N. Staroverov, R. Kobayashi, J. Normand, K. Raghavachari, A. Rendell, J. C. Burant, S. S. Iyengar, J. Tomasi, M. Cossi, N. Rega, N. J. Millam, M. Klene, J. E. Knox, J. B. Cross, V. Bakken, C. Adamo, J. Jaramillo, R. Gomperts, R. E. Stratmann, O. Yazyev, A. J. Austin, R. Cammi, C. Pomelli, J. W. Ochterski, R. W. Martin, K. Morokuma, V. G. Zakrzewski, G. A. Voth, P. Salvador, J. J. Dannenberg, S. Dapprich, A. D. Daniels, O. Farkas, J. B. Foresman, J. V. Ortiz, J. Cioslowski and D. J. Fox, *Gaussian 09, Revis. D.01*, 2009.

Chapter 3

Connecting Mononuclear Dysprosium Single-Molecule Magnets to Form Dinuclear Complexes *via in situ* Ligand Oxidation

3.1 Introduction

Since the discovery of the magnet-like behaviour of slow relaxation of the magnetization in individual molecules, single-molecule magnets (SMMs) have emerged as prominent systems in regard to molecular magnetism.[1] This is primarily due to the fact that these molecules hold tremendous potential towards implementation in quantum computation and memory storage devices.[2] Over the years, several hundreds of research groups have focused their attention on the synthesis of new SMMs with large energy barriers using various anisotropic $3d$ metal ions.[3–7] However, the recent development of lanthanides as ideal candidates for the isolation of high barrier SMMs has shifted synthetic efforts due to the considerable intrinsic magnetic anisotropy of the late $4f$ elements.[8–11]

Although lanthanide-SMMs hold the current record for large anisotropy barriers, significant improvements are necessary for their realization in the aforementioned applications. For instance, inducing magnetic coupling between two $4f$ ions remains a major challenge due to the core nature of the $4f$ orbitals. In addition, controlling the symmetry, coordination geometry and the crystal field effects are not trivial. This therefore necessitates the design, synthesis and study of novel systems. When it comes to developing new SMMs, ligand design is critical. To that end, ligands must be efficient chelates while also harnessing the intrinsic anisotropy of lanthanide ions.[12–14] Ligands that can also act as bridging units between metal centres are particularly attractive as they can also provide potential magnetic exchange pathways often termed super-exchange. In most cases, ligands are carefully chosen for these chelating and/or bridging properties. In that regard, 2,2';6',2''-terpyridine (terpy; Chart 3.1) is a well-known tridentate chelating ligand that has been used for several decades for encapsulating metal ions. However, terpy does not provide any possibilities for bridging metal ions. In contrast, 3,5-bis(2-pyridyl)-4H-1,2,4,6-thiatriazine (Py$_2$TTAH; Chart 3.1) is an inherently redox-active pincer-type ligand as discussed in Chapter 2. The versatility of Py$_2$TTAH as a ligand is shown upon deprotonation to act as an anionic unit (Py$_2$TTA$^-$) or subsequent oxidation to generate the neutral radical (Py$_2$TTA), which can then be further oxidized to form 1-oxo-3,5-bis(2-pyridyl)-4-hydro-1,2,4,6-thiatriazine (1-oxo-Py$_2$TTAH); all while retaining its stable coordination pocket (Chart 3.1).[15,16] Therefore, Py$_2$TTAH is an ideal ligand system to explore in the coordination chemistry of lanthanide ions.

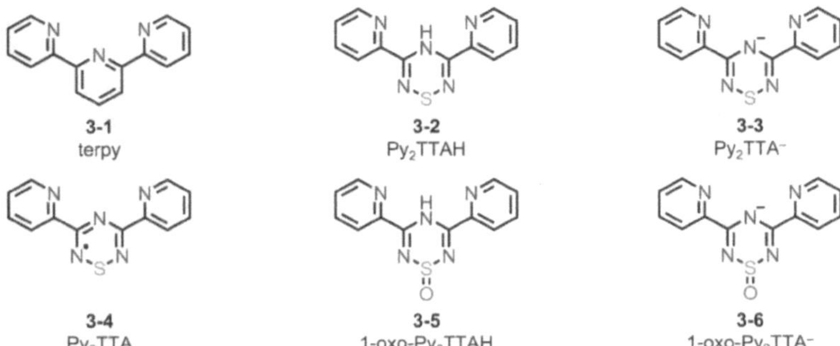

Chart 3.1 Thiatriazinyl-based ligands as structural mimics to 2,2';6',2''-terpyridine.

The synthesis of SMMs based on coordination chemistry relies on a self-assembly process. In this approach, energetically favourable coordination complexes are formed and subsequently studied through single-crystal X-ray diffraction and magnetometry in order to establish magneto-structural correlations. Little importance is typically given to the self-assembly process as the reaction intermediates are often hard to isolate or study since these reactions involve several ingredients in a one-pot synthesis. However, if the self-assembly process can be monitored and studied, invaluable information on the complex formation can be achieved. This can subsequently be applied to the synthesis of targeted molecules. With this in mind, we have focused our attention on the synthesis of Py_2TTA-based lanthanide SMMs, not only for their magnetic behaviour but also to investigate their self-assembly process. Herein we report the synthesis, isolation and study of $[Dy_2(1\text{-}oxo\text{-}Py_2TTA)_2(NO_3)_4(H_2O)_2]\cdot CH_3CN$ (**3-7a**), exhibiting SMM properties. In addition, its yttrium analogue (**3-7b**) is reported in order to elucidate the self-assembly of our SMM *via* NMR studies.

3.2 Results and Discussion

3.2.1 Synthesis, Crystallography and Time Dependent ^{1}H-NMR Studies

Reaction of $Dy(NO_3)_3\cdot6H_2O$ (1 equiv.) with **3-2** (1 equiv.) in acetonitrile (MeCN) yielded a dark purple solution which turns clear colourless within one hour (Scheme 3.1). After 3 hours, the formation of pale yellow rectangular shaped crystals of **3-7a**·MeCN resulted in 60% yield. The drastic change in colour of the reaction mixture indicates a change in the "redox" chemistry of the ligand upon coordination. This is further evidenced by the single-crystal X-ray structure, which indicates that metal assisted oxidation of the ligand occurs (*vide infra*). It is noteworthy that an increase in metal concentration (e.g., from 1 to 2 equivalents) in the reaction mixture accelerates the rate of the change in colour. In addition, the size and shape of the reaction vessel influence the rate of the reaction. Based on these observations, it is clear that some interesting redox chemistry is occurring, as Py_2TTAH is surprisingly robust with a high tolerance towards aqueous, basic, and thermal conditions. Its only instability appears to be under acidic conditions, in which the thiatriazine (TTA) ring opens and regenerates an *N*-bridgehead-1,2,5-thiadiazolium salt.[17] Furthermore, oxidation of Py$_2$TTAH in the absence of metals is not achievable. For example, bubbling O$_2$ gas through solutions of Py$_2$TTAH in methanol, MeCN (wet and dry) and dichloroethane left Py$_2$TTAH intact (i.e., no oxidation occurred). If, however, the ligand is coordinated to a metal ion, oxidation becomes more feasible, as has been observed in previous coordination studies employing this ligand system.[15,16]

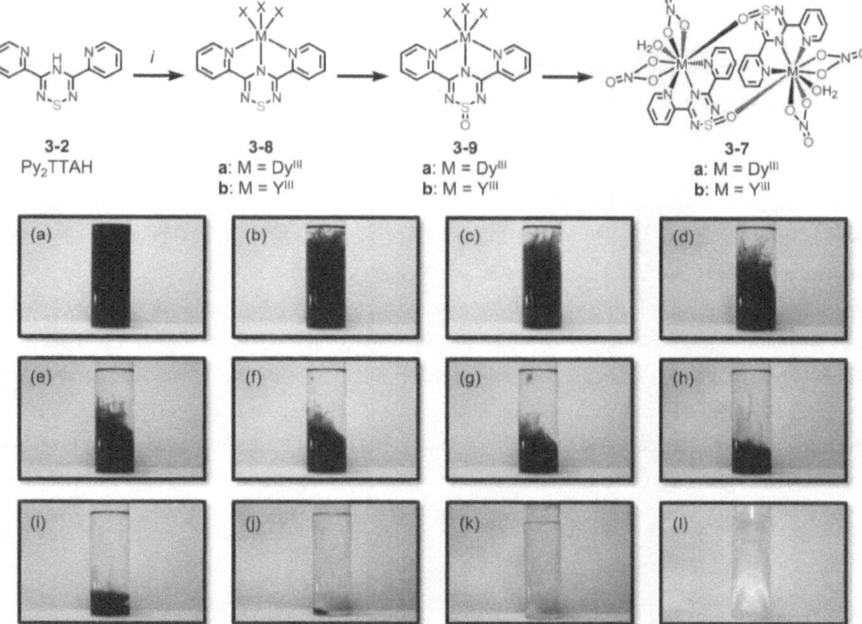

Scheme 3.1 (Top) Synthesis of complexes **3-7a** and **3-7b**. X = NO$_3$ or H$_2$O. Reagents and conditions: (*i*) M(NO$_3$)$_3$·6H$_2$O, MeCN, 25 °C. (Bottom) Progression of the reaction mixture at *t* = (a) 0 min, (b–k) 80–100 min, (l) 3 h.

Since Py$_2$TTA can exist as a neutral radical in solution, the generation of the oxide may be attributed to the generation of the radical, which is known to be susceptible to oxidation.[18] Alternatively, the generation of the oxide may occur following coordination to the metal ion, which pulls electron density of the anionic charge towards the metal centre leaving the sulfur atom of the TTA ring susceptible to oxidation. In order to elucidate the reaction process, time dependent ^{1}H NMR studies were carried out on the *in situ*

53

reaction between Y(NO$_3$)$_3$·6H$_2$O (2 equiv.) and **3-2** (1 equiv.) in CD$_3$CN. The choice of yttrium salt over Dy(NO$_3$)$_3$·6H$_2$O for the NMR studies was due to the diamagnetic nature, which would enable the determination of a radical intermediate (e.g., if a paramagnetic signal was observed this could be attributed to the generation of the Py$_2$TTA radical). Conversely, if diamagnetic intermediates are formed, using yttrium in place of dysprosium enables identification of the intermediates *via* NMR spectroscopy. In order to correlate these results with the reaction observed between Dy(NO$_3$)$_3$·6H$_2$O and **3-2**, in addition to similar physical properties of the reaction (i.e., bleaching of the reaction mixture), X-ray analysis confirmed the generation of isostructural [Y$_2$(1-oxo-Py2TTA)$_2$(NO$_3$)$_4$(H$_2$O)$_2$]·CH$_3$CN (**3-7b**·CH$_3$CN; *vide infra*).

Time dependent ^{1}H NMR studies, which are summarized in Figure 3.1, reveal that the reaction mixture initially consists solely of the unreacted Py$_2$TTAH starting material. After 15–20 min, four additional pyridyl peaks begin to emerge, which is attributed to the generation of a yttrium coordinated Py$_2$TTA$^-$ ligand intermediate (i.e., **3-8b**). Within 5–10 min of the observation of the first intermediate, a second compound also exhibiting a pattern consistent with four pyridyl peaks appears, which is postulated to be the oxidized anionic ligand, 3,5-bis(pyridyl)-1-oxo-1,2,4,6-thiatriazide (1-oxo-Py$_2$TTA$^-$) coordinated to yttrium (i.e., **3-9b**). After 60 min, **3-2**, **3-8b** and **3-9b** are all present in the solution. After 90 min, we begin to see the depletion of Py$_2$TTAH and broadening of the NMR peaks, which is attributed to precipitation of the dimer within the NMR tube. Following the complete consumption of **3-2** (6 h spectrum), the N–H peaks no longer exist, providing evidence for these assignments as metal coordinated ligands. Comparison of the peak shifts with that of unreacted Py$_2$TTAH and the 1-oxo-Py$_2$TTAH (prepared *via* oxidation of the radical), coupled with the formation of the dimer (which precipitates out of solution), further supports these assignments.

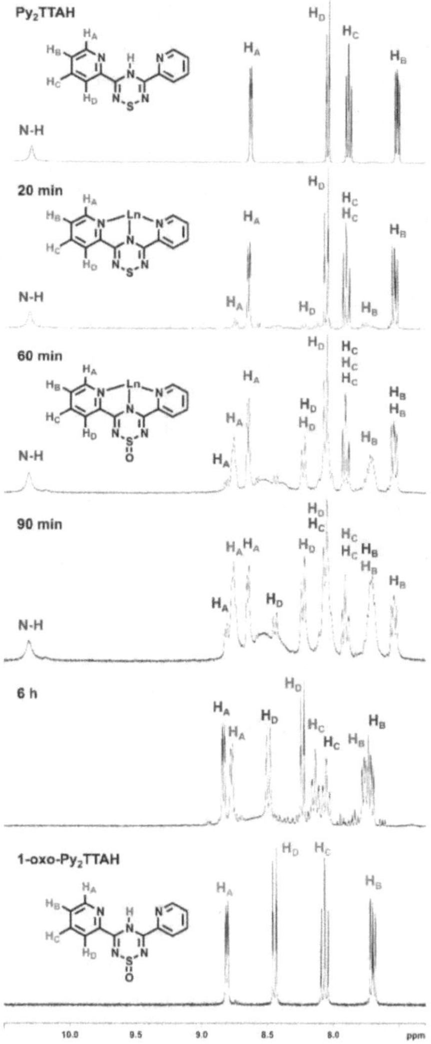

Figure 3.1 Time-dependent ¹H NMR study on the *in situ* reaction between Y(NO₃)₃·6H₂O (2 equiv.) and **3-2** (1 equiv.) in CD₃CN (1 mL, 6.4 x 10⁻² M).

Due to the isostructural nature of complexes **3-7a** and **3-7b**, only the dysprosium analogue is discussed below. The asymmetric unit of the centrosymmetric structure of **3-7a** is composed of a Dy^{III} ion with a nine-coordinate environment adopting a distorted muffin geometry (Figure 3.2).[19] The tridentate terpyridine-like coordination pocket of the 1-oxo-Py$_2$TTA$^-$ ligand binds to the metal centre *via* N1, N4 and N5 donors while oxygen atoms (O2, O3, O5, O6) from two nitrate anions and a water molecule (O8) occupy the primary coordination environment. The remaining single coordination site is occupied by an oxygen atom (O1a) from the oxidized S=O group of the neighbouring 1-oxo-Py$_2$TTA$^-$ ligand. This S=O group plays an integral role in the dimerization and formation of the dinuclear molecule. The intramolecular Dy$\cdots$Dy distance of 7.09 Å is relatively long compared to the reported dinuclear complexes; however, weak intramolecular magnetic interactions cannot be ignored for highly anisotropic Dy^{III} ions.[20] It is interesting to note that there are no short contacts or hydrogen bonding between the asymmetric units; only the S=O bridges hold the dinuclear molecules together. Close inspection of the packing arrangement reveals that the molecules are well aligned with each other with the closest inter-metal distance being 6.76 Å (Figure 3.3)

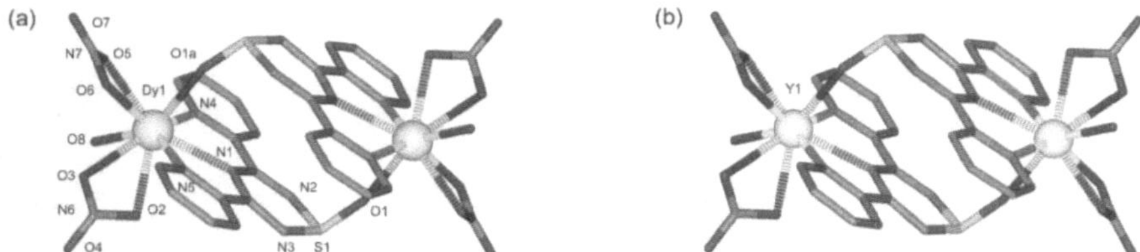

Figure 3.2 Partially labelled molecular X-ray structure of the centrosymmetric dinuclear complexes (a) **3-7a**·CH$_3$CN and (b) **3-7b**·CH$_3$CN. Hydrogen atoms and MeCN molecules are omitted for clarity.

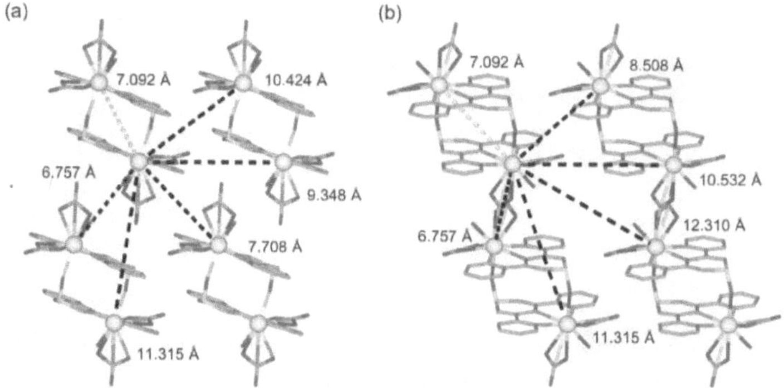

Figure 3.3 Crystallographic packing of **3-7a**·MeCN down the (a) *b*-axis and (b) *a*-axis. Dashed lines infer intra- (peach) and intermolecular (black) Dy···Dy distances. Hydrogen atoms and MeCN molecules are omitted for clarity.

3.2.2 Magnetic Measurements

In order to elucidate the magnetic properties of the paramagnetic complex **3-7a**, direct current (DC) and (AC) magnetic susceptibility measurements were performed using a superconducting quantum interference device (SQUID) magnetometer. The DC susceptibility measurements at 1000 Oe (Figure 3.4) reveal a room temperature χT value of 27.97 cm^3 K mol^{-1} consistent with the expected value of 28.34 cm^3 K mol^{-1} for two non-interacting Dy^{III} ions ($S = 5/2$, $L = 5$, $^6H_{15/2}$, $g = 4/3$). Upon lowering the temperature, the χT product decreases gradually to reach a minimum value of 18.80 cm^3 K mol^{-1} at 1.8 K. Such a negative deviation suggests the existence of weak antiferromagnetic interactions between metal ions and/or the presence of magnetic anisotropy. Intra- and intermolecular distances (7.09 and 6.76 Å, respectively) suggest that at this applied field interactions between the ions are likely rather small, if not negligible, due to the core nature of the 4*f* ions.[21] Therefore, it is reasonable to attribute the decrease of the χT product to the magnetic anisotropy of the Dy^{III} ions.

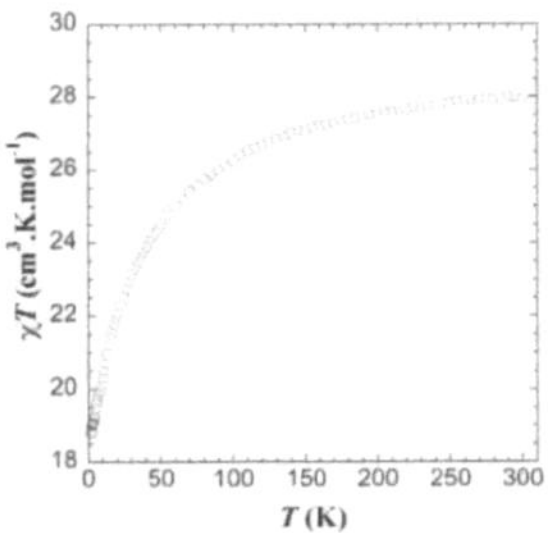

Figure 3.4 Temperature dependence of the χT product at 1000 Oe for **3-7a**·MeCN with χ being the molar magnetic susceptibility equal to M/H per mole of the compound.

Recent studies have shown a nine-coordinate environment around the Dy^{III} ion that can provide non-negligible magnetic anisotropy to a molecular system if an ideal ligand field is present.[19] Here, the two nitrate ions around the metal centre play an important electron withdrawing effect by promoting higher degeneracy of the $4f$ orbitals subsequently leading to enhanced spin-orbit coupling.[22] In order to further confirm the magnetic anisotropy, magnetization measurements were conducted (Figure 3.5). The magnetization plot M vs. H/T shows field dependence of the magnetization that does not saturate at low temperatures and high magnetic fields indicating the presence of magnetic anisotropy and/or low-lying excited states in this system, while the non-superimposition of the isothermal lines is an indication of the former.

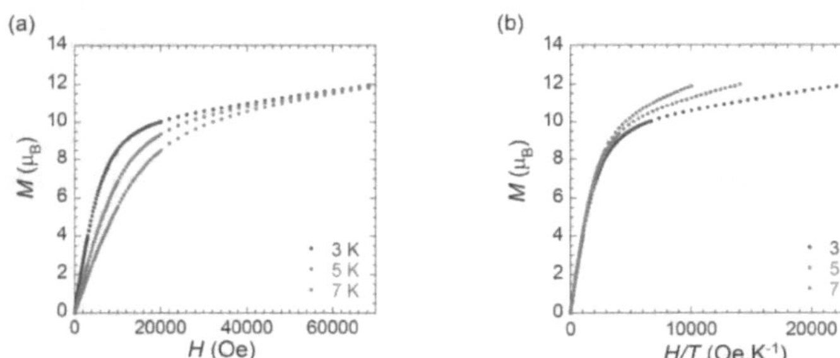

Figure 3.5 Field dependence of the magnetization, M, at the indicated temperatures (a) for **3-7a**·MeCN illustrating the non-saturation. M vs. H/T plot (b) showing non-superposition of the magnetization curves at different temperatures.

To further explore the possibility of SMM behaviour, AC magnetic susceptibility measurements were carried out under a zero DC field. The characteristic frequency dependent out-of-phase signal was not observed; this can be attributed to significant quantum tunnelling of the magnetization (QTM) often observed in lanthanide SMMs, which diminishes the relaxation barriers. Such QTM can be reduced upon application of a small static field. In order to determine the optimum DC field, AC susceptibility measurements at various static fields (250–3000 Oe) were performed at 2 K (Figure 3.6a,b). The out-of-phase data clearly exhibit an optimum DC field (the field at which the minimum of characteristic frequency is observed, Figure 3.6c) of 750 Oe, where the QTM is minimized. AC measurements at this optimum field reveal a full frequency dependent peak below 5 K between 0.1 and 1000 Hz (Figure 3.6e,f) indicative of SMM behaviour. Analysis of the data using the Arrhenius law (Equation 3.1) gave a calculated relaxation barrier of U_{eff} = 19 K and a τ_0 value of 2.0 x 10^{-7} s for the thermally activated regime (Figure 3.6d).

$$\tau = \tau_0 e^{U_{eff}/k_B T}$$

(3.1)

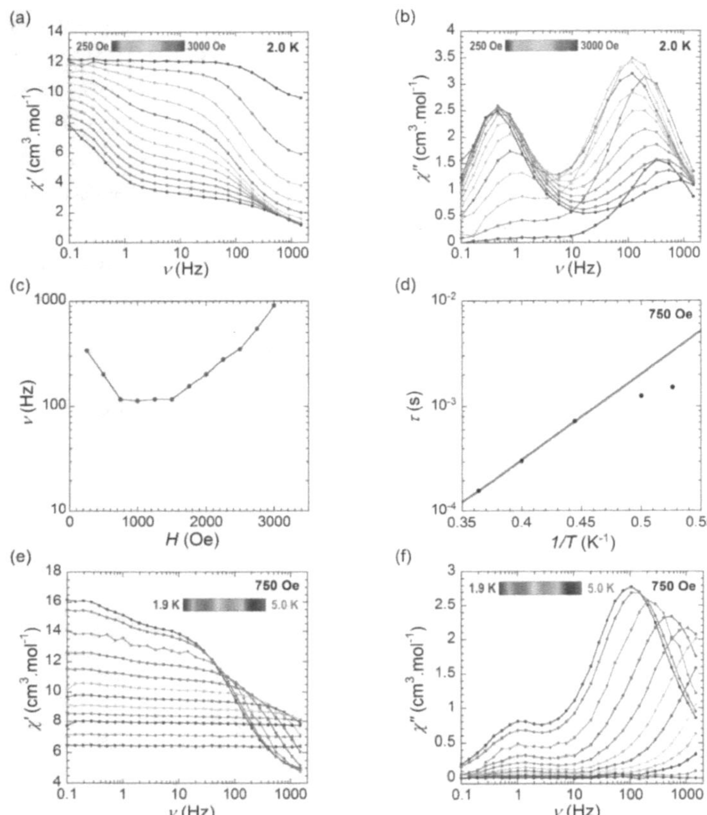

Figure 3.6 Frequency dependence of the (a) in-phase χ' and (b) out-of-phase χ'' AC susceptibility signals under various DC fields at 2 K; (c) field dependence of the characteristic frequency as a function of the applied DC field at 2 K; (d) magnetization relaxation time τ versus $1/T$ plot under the optimum field of 750 Oe (red line corresponds to Arrhenius law) and frequency dependence of the (e) in-phase χ' and (f) out-of-phase χ'' AC susceptibility signals under 750 Oe DC field for **3-7a**·MeCN at indicated temperatures.

The observed barrier is relatively small as a result of the QTM. It is noteworthy that the presence of a peak shoulder around 1 Hz can be observed in Figure 3.6f. The intensity of the peak shoulder grows to become a full peak when the applied static field is increased to 2500 Oe and diminishes at smaller fields such as 250 Oe (Figure 3.6b). Moreover, AC data collected at the applied static field of 2500 Oe exhibit this peak around 1 Hz, which does not shift (Figure 3.7). Such behaviour can arise from large field-induced intermolecular interactions rather than arising from the molecular origin.[23] This is also not completely surprising given the fact that the closest inter-metal distance of 6.76 Å is shorter than the intramolecular Dy···Dy distance.

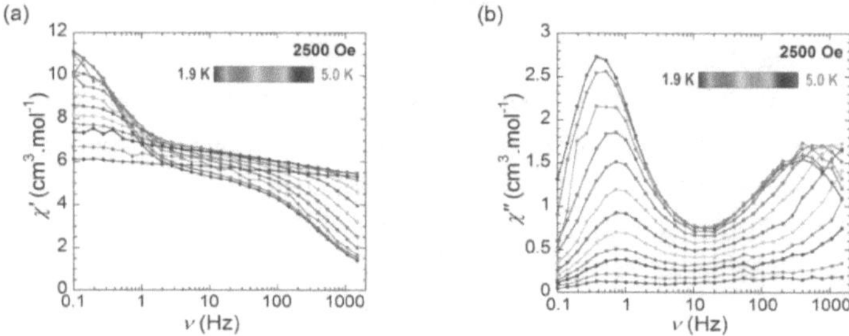

Figure 3.7 Frequency dependence of the (a) in-phase χ' and (b) out-of-phase χ'' AC susceptibility signals under 750 Oe DC field for **3-7a**·MeCN at indicated temperatures.

3.3 Summary and Conclusion

In conclusion, we have successfully isolated a dinuclear dysprosium complex and its yttrium analogue *via in situ* metal-assisted ligand oxidation of our pincer type Py$_2$TTA followed by dimerization. This unique metal complexation was carefully probed through solution NMR studies, which clearly demonstrates the ability of the Py$_2$TTA$^-$ ligand to oxidize and form the 1-oxo-Py$_2$TTA$^-$ upon metal coordination. Subsequently, the oxidized ligand coordinates to the oxophilic lanthanide ion thus forming discrete dinuclear molecules.

The paramagnetic Dy^{III} analogue displays field-induced single-molecule magnet behaviour where the origin is attributed to the single-ion anisotropy of the dysprosium ion. Although **3-7a** is structurally a dinuclear molecule, due to the large intramolecular separation, it acts as a mononuclear SMM. Higher applied DC fields during the AC measurements indicate the presence of a secondary relaxation process that can be attributed to intermolecular interactions rather than being molecular in origin. This study provides a methodology for probing/visualizing reactions that can be used to understand the nature of self-assemblies and their driving forces.

3.4 Experimental Details

3.4.1 General Procedures

The reagents dysprosium (III) nitrate hexahydrate and yttrium (III) nitrate hexahydrate (Strem) were obtained commercially and used as received. Acetonitrile was of reagent grade. The ligand 3,5- bis(2-

pyridyl)-4H-1,2,4,6-thiatriazine (**3-2**) was synthesized according to literature procedures.[17] Melting points were taken using a Mel-Temp apparatus and are uncorrected. NMR spectra were run in CD_3CN solutions at room temperature on a Bruker Avance II 300 MHz spectrometer. FTIR spectra of solid samples were recorded on an Agilent Technologies Cary 630 FTIR spectrometer.

3.4.2 Experimental Procedures

Preparation of [Dy$_2$(1-oxo-Py$_2$TTA)$_2$(NO$_3$)$_4$(H$_2$O)$_2$]·CH$_3$CN (3-7a·MeCN). $Dy(NO_3)_3 \cdot 6H_2O$ (85 mg, 0.254 mmol) and **3-2** (66.7 mg, 0.261 mmol) were dissolved in acetonitrile (40 mL) to create a purple solution and was stored at RT. A colourless solution emerged after 1 h with crystallization of a white solid. Yield = 60%. dec >260°C. IR (υ_{max}, ATR): 3332 (br), 3248 (br), 3117 (w), 3100 (w), 1647 (w), 1611 (w), 1600 (w), 1584 (w), 1559 (m), 1498 (s), 1482 (s), 1445 (m), 1407 (m), 1357 (s), 1302 (m), 1281 (s), 1254 (m), 1208 (w), 1162 (w), 1153 (w), 1112 (w) 1102 (w). 1094 (w), 1057 (m), 1036 (s), 1027 (s), 1016 (s), 1008 (s), 977 (w), 912 (w), 901 (w), 815 (m), 809 (m), 787 (m), 759 (s), 751 (s), 748 (s), 736 (m), 724 (s), 709 (m), 697 (m) cm^{-1}.

Preparation of [Y$_2$(1-oxo-Py$_2$TTA)$_2$(NO$_3$)$_4$(H$_2$O)$_2$]·CH$_3$CN (3-7b·MeCN). The synthetic procedure followed is the same as that mentioned above using $Y(NO_3)_3 \cdot 6H_2O$ (68 mg, 0.249 mmol) and **3-2** (64.0 mg, 0.251 mmol). Yield = 65%. dec >260°C. IR (υ_{max}, ATR): 3335 (br), 3261 (br), 3120 (w), 3103 (w), 1649 (w), 1612 (w), 1600 (w), 1584 (w), 1559 (m), 1502 (s), 1482 (m), 1445 (m), 1407 (m), 1360 (s), 1303 (m), 1285 (s), 1254 (m), 1208 (w), 1162 (w), 1153 (w), 1113 (w), 1102 (w), 1095 (w), 1058 (m), 1038 (s), 1029 (s), 1017 (s), 1009 (s), 977 (w), 912 (w), 901 (w), 815 (m), 810 (m), 788 (m), 759 (s), 753 (s), 749 (s), 736 (m), 725 (s), 709 (w), 697 (m).

3.4.3 Crystallography

Colourless blocks of single crystals of complex **3-7a,b**·MeCN were obtained and collected. Crystallographic data as well as data collection and refinement are summarized in Table A1.5 and selected bond distances are given in Table A1.6 of Appendix A1. Single-crystal X-ray diffraction data collection was performed at 200 K using a Bruker APEX II CCD detector with a sealed Mo tube source (wavelength 0.71073 Å). Raw data collection and processing were performed with APEX II software package from Bruker AXS. Initial unit cell parameters were determined from 60 data frames with 0.3° ω scan each, collected at the different sections of the Ewald sphere. Semi-empirical absorption corrections based on equivalent reflections were applied.[24] The structure was solved by direct methods, completed with difference Fourier synthesis, and refined with full-matrix least-squared procedures based on F^2. All non-hydrogen atoms were refined with anisotropic thermal motion approximation. All hydrogen atom positions were calculated based on the geometry of related non-hydrogen atoms. All hydrogen atoms were treated as idealized contributions during the refinement. All scattering factors are contained in several versions of the SHELXTL program library, with the latest version used being v.6.12.[25]

3.4.4 Magnetic Measurements

The magnetic susceptibility measurements were obtained using a Quantum Design SQUID magnetometer MPMS-XL7 operating between 1.8 and 300 K. DC measurements were performed on a polycrystalline sample of complex **3-7a**·MeCN of 16.3 mg. The sample was wrapped in a polyethylene membrane and subjected to fields in a range from 0 to 7 T. The magnetization data was collected at 100 K in order to check for ferromagnetic impurities that were found to be absent in the sample. Diamagnetic corrections were applied to correct for contribution from the sample holder, and the inherent diamagnetism of the sample was estimated with the use of Pascal's constants.

3.5 References

1   G. Christou, D. Gatteschi, D. N. Hendrickson and R. Sessoli, *MRS Bull.*, 2000, **25**, 66–71.

2   L. Bogani and W. Wernsdorfer, *Nat. Mater.*, 2008, **7**, 179–186.

3   T. C. Stamatatos, D. Foguet-Albiol, C. C. Stoumpos, C. P. Raptopoulou, A. Terzis, W. Wernsdorfer, S. P. Perlepes and G. Christou, *J. Am. Chem. Soc.*, 2005, **127**, 15380–15381.

4   C. J. Milios, A. Vinslava, W. Wernsdorfer, S. Moggach, S. Parsons, S. P. Perlepes, G. Christou and E. K. Brechin, *J. Am. Chem. Soc.*, 2007, **129**, 2754–2755.

5   G. Aromí and E. K. Brechin, in *Single-Molecule Magnets and Related Phenomena*, ed. R. Winpenny, Springer Berlin Heidelberg, 2006, vol. 122, pp. 1–67.

6   M. Murrie, *Chem. Soc. Rev.*, 2010, **39**, 1986–1995.

7   C. Lampropoulos, C. Koo, S. O. Hill, K. Abboud and G. Christou, *Inorg. Chem.*, 2008, **47**, 11180–11190.

8   J. Tang, I. Hewitt, N. T. Madhu, G. Chastanet, W. Wernsdorfer, C. E. Anson, C. Benelli, R. Sessoli and A. K. Powell, *Angew. Chemie Int. Ed.*, 2006, **45**, 1729–1733.

9   N. Ishikawa, M. Sugita, T. Ishikawa, S. Y. Koshihara and Y. Kaizu, *J. Am. Chem. Soc.*, 2003, **125**, 8694–8695.

10   P.-H. Lin, T. J. Burchell, R. Clérac and M. Murugesu, *Angew. Chemie Int. Ed.*, 2008, **47**,

8848–8851.

11   S. Cardona-Serra, J. M. Clemente-Juan, E. Coronado, A. Gaita-Ariño, A. Camón, M. Evangelisti, F. Luis, M. J. Martínez-Pérez and J. Sesé, *J. Am. Chem. Soc.*, 2012, **134**, 14982–14990.

12   J. D. Rinehart and J. R. Long, *Chem. Sci.*, 2011, **2**, 2078.

13   C. Aronica, G. Pilet, G. Chastanet, W. Wernsdorfer, J. F. Jacquot and D. Luneau, *Angew. Chemie - Int. Ed.*, 2006, **45**, 4659–4662.

14   P. H. Lin, W. Bin Sun, M. F. Yu, G. M. Li, P. F. Yan and M. Murugesu, *Chem. Commun.*, 2011, **47**, 10993–10995.

15   K. L. M. Harriman, A. A. Leitch, S. A. Stoian, F. Habib, J. L. Kneebone, S. I. Gorelsky, I. Korobkov, S. Desgreniers, M. L. Neidig, S. Hill, M. Murugesu and J. L. Brusso, *Dalt. Trans.*, 2015, **44**, 10516–10523.

16   K. L. M. Harriman, I. A. Kühne, A. A. Leitch, I. Korobkov, R. Clérac, M. Murugesu and J. L. Brusso, *Inorg. Chem.*, 2016, **55**, 5375–5383.

17   A. A. Leitch, I. Korobkov, A. Assoud and J. L. Brusso, *Chem. Commun.*, 2014, **50**, 4934–4936.

18   N. J. Yutronkie, A. A. Leitch, I. Korobkov and J. L. Brusso, *Cryst. Growth Des.*, 2015, **15**, 2524–2532.

19   A. K. Mondal, S. Goswami and S. Konar, *Dalt. Trans.*, 2015, **44**, 5086–5094.

20   F. Habib and M. Murugesu, *Chem. Soc. Rev.*, 2013, **42**, 3278.

21   R. A. Layfield, J. J. W. McDouall, S. A. Sulway, F. Tuna, D. Collison and R. E. P. Winpenny, *Chem. - A Eur. J.*, 2010, **16**, 4442–4446.

22   F. Habib, G. Brunet, V. Vieru, I. Korobkov, L. F. Chibotaru and M. Murugesu, *J. Am. Chem. Soc.*, 2013, **135**, 13242–13245.

23   F. Habib, I. Korobkov and M. Murugesu, *Dalt. Trans.*, 2015, **44**, 6368–6373.

24   R. H. Blessing, *Acta Crystallogr. Sect. A*, 1995, **51**, 33–38.

25   G. M. Sheldrick, *Acta Crystallogr. Sect. A Found. Crystallogr.*, 2008, **64**, 112–122.

Chapter 4

Molecular Influences Towards the Solid-State Packing of Pyridine-Bridged Bisthiadiazinyl Radicals

4.1 Introduction

By virtue of their open-shell structure, molecular radicals inherently possess a conductive and magnetic nature that have led to their recognition towards the development of future devices.[1–10] Integration of radicals as building blocks for these applications remains an ongoing feat; that is, designing a molecular template that not only stabilizes the intrinsic instability of organic radicals but enables control over its properties with precision.[1,11,12] Part of this control entails the nanostructuring of the self-assembly of molecules and the degree to which radicals interact with one another, so as to install unique electronic and magnetic functionality. For synthetic chemists, physicists and engineers alike, understanding the relationship between the solid state and physicochemical properties of open-shell materials continues to be a growing interest, in which new advancements can be made through the structural versatility of organic-based molecules. Regarding this matter, heterocyclic thiazyl radicals have been pursued by researchers for their desirable physical attributes. Inclusion of the thiazyl (S–N) unit within molecular π-frameworks can provide a landscape for stable electronic configurations and, in addition, be a source of favourable π–π and electrostatic interactions (e.g., $S\cdots N'$, $S\cdots S'$) towards the construction of supramolecular architectures. As these interactions in many cases are linked to the spin of the unpaired electron, pathways for electronic and magnetic communication are often sensitive to structural changes. Accordingly, the solid-state properties of heterocyclic thiazyls can be fine-tuned through synthetic modifications in line with the peripheral exocyclic groups,[13] from which numerous thiazyl radicals have been explored as molecular conductors,[14,15] magnets[16–19] and switches[20] over the decades. Many of these examples are found within the families of dithiadiazolyls[21] and bisdithiazolyls[15,22] (Chart 4.1; DTDAs and bis-DTAs, respectively) that has encompassed a history of rich structural diversity. However, classes of thiazyls including thiatriazinyls[23–25] and bisthiadiazinyls[26,27] (Chart 4.1; TTAs and bis-TDAs, respectively) are few and far between. From this perspective, opportunities to explore the structure-property relationship of the latter await with judicious choice of substituents.

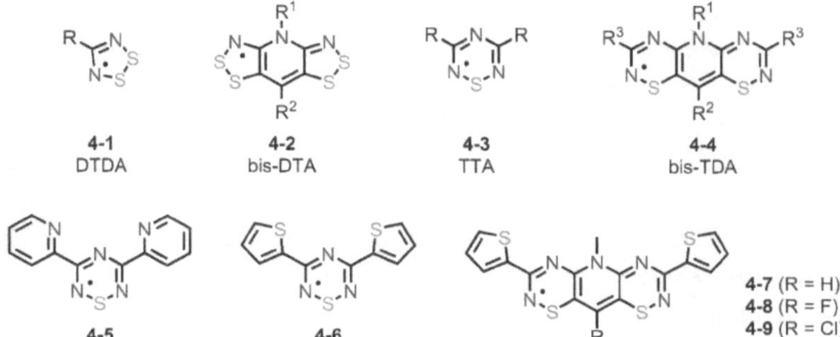

Chart 4.1 Examples of monocyclic and resonance-stabilized thiazyl radicals.

In recognition of this, we recently reported the synthesis and characterization of heteroaromatic-substituted TTAs with 2-pyridyl and 2-thienyl attachments (**4-5** and **4-6** in Chart 4.1).[28,29] Differences between the two heteroaromatic moieties, in terms of both chemical nature and composition, were shown to influence the spin distribution across the molecular framework, while providing interactions fruitful of directing the assembly of molecules in the crystalline phase. Aware of the potential behind heteroaromatic functionalization, we have continued to explore the influences of the heterocyclic appendages to the bis-TDA frameworks and herein discuss the synthesis and characterization of the thienyl-substituted pyridine-bridged bisthiadiazinyl **4-7** (R = H). Moreover, synthetic routes towards fluorine and chlorine substitution at the basal carbon position have been discovered, from which we also report the preparation of the fluoro- (**4-8**, R = F) and chloro-analogues (**4-9**, R = Cl). Spectroscopic, electrochemical and computational studies have revealed both thienyl extension and halogenation to be effective methods for fine-tuning the electronic properties of the bis-TDA framework. Furthermore, solid-state analysis has unveiled temperature-dependent polymorphism to exist for all three radicals with at least five phases observed for the proto-analogue. The range of supramolecular structures highlights a degree of flexibility in guiding the crystal packing brought forth by the thienyl moieties and crystallographic effects upon introducing halogens to the molecular system. Fundamentally, this work provides an avenue towards functionalization of resonance-stabilized bis-TDA radicals and insights towards electronic and crystal engineering of thiazyl-based materials.

4.2 Results and Discussion

4.2.1 Synthesis, Optical Spectroscopy and Electrochemistry of bis-TDA Cations

Inspired by the synthesis of related pyridine-bridged stable radicals,[30] isolation of the thienyl-functionalized bis-TDAs was achieved following the route outlined in Scheme 4.1. For the proto- derivative

4-7, this began with the alkylation of 2,6-difluoropyridine (**4-10**) with methyl triflate (MeOTf) to afford *N*-methyl-2,6-difluoropyridinium triflate (**4-12**). Subsequent treatment of an acetonitrile (MeCN) solution of **4-12** with an excess of 2-thiophenecarboxamidine in MeCN led to the precipitation of amidine hydrofluoride. Upon filtration and removal of the solvent from the filtrate, the pyridinium-bridged bisamidine **4-14** and triamidine by-product **4-16** are isolated as a yellow orange mixture. Identification of **4-16** was confirmed through SCXRD, in which anion metathesis can be achieved by reacting the crude triflate salt with benzyltriethylammonium chloride followed by recrystallization in isopropanol to afford crystals of **4-17** (Figure 4.1a). Alternatively, facile separation of the initial mixture of triflate salts can be realized *via* recrystallization from hot water, affording yellow needles of pure **4-14**. Following the addition of four equivalents of sulfur monochloride (S_2Cl_2) to a chlorobenzene solution of **4-14** under gentle reflux, the closed shell bis-TDA cation **4-18** can be isolated as a red microcrystalline solid. It is worth noting, chlorination at the basal position can occur when the double Herz cyclocondensation of **4-14** is carried out in refluxing MeCN, leading to a mixture of the desired product **4-18** along with **4-20**; however, this can be avoided by using chlorobenzene. Recrystallization from hot MeCN affords **4-18** as ruby coloured crystals.

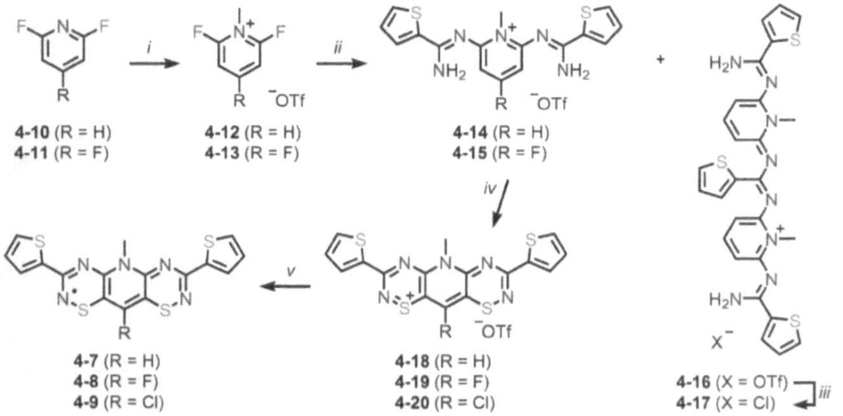

Scheme 4.1 Synthesis of bis-TDA radicals 4-7–4-9. Reagents: (*i*) MeOTf, MeCN; (*ii*) 2-thiophenecarboxamidine, MeCN; (*iii*) triethylbenzylammonium chloride, MeCN; (*iv*) S_2Cl_2, MeCN; (*v*) DiMFc, MeCN.

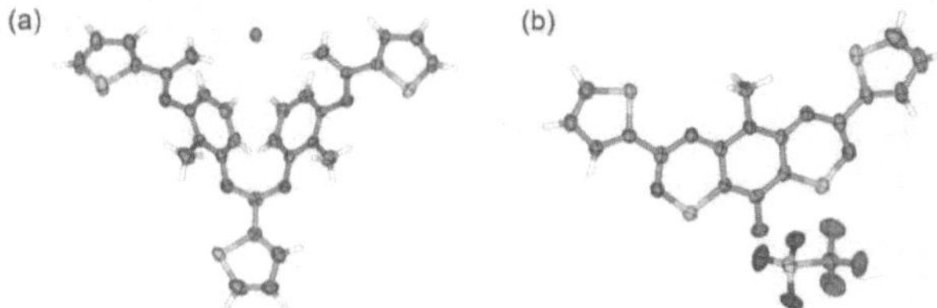

Figure 4.1 ORTEP drawings (50% thermal ellipsoids) of the asymmetric unit of (a) **4-17** and (b) **4-19** at 200 K (for crystallographic data see Tables A1.7 in Appendix A1). Disorder in the thiophene moieties is not shown. Colour code: C, grey; H, white; N, blue; O, Red; F, pink; S, yellow; Cl, green.

Expanding this study to explore the influence of the basal substituent on the structure-property relationships was achieved through halogenation at the R position of the bis-TDA radicals (i.e., **4-8** and **4-9**). With respect to the fluoro-derivative, synthesis of **4-19** follows an analogous procedure to its proto-counterpart, although starting from 2,4,6-trifluoropyridine (**4-11**). Interestingly, upon undergoing the double Herz cyclocondensation, the ideal solvent system for the fluoro-derivative was determined to be a 19:1 mixture of dichloroethane (DCE) and MeCN. Such meticulous attention to the reaction conditions was necessary in order to prevent substitution of the fluorine atom for chlorine, which was observed as a minor by-product when 100% MeCN was used. Upon recrystallization from hot MeCN, red block-like needles of **4-19** suitable for single-crystal X-ray analysis were obtained, confirming its identity (Figure 4.1b).

Based on the observation of **4-20** as a by-product in the formation of **4-18**, we focused our attention on altering the reaction conditions to promote chlorination at the basal position, rather than using 4-chloro-2,6-difluoropyridine as a precursor. Specifically, employing 100% MeCN as the solvent, increasing the number of equivalents of S_2Cl_2 to eight (*cf.* four for **4-18** and **4-19**) and lengthening the reaction time to 16h (*cf.* 2h for **4-18** and **4-19**), complete chlorination of the central carbon atom was achieved, affording **4-20** as a red microcrystalline solid. Recrystallization in MeCN yielded the product as red needles.

UV-Vis spectroscopic analysis of bis-TDA cations **4-18**–**4-20** was completed, the results of which are presented in Figure 4.2a and Table 4.1. Although the recrystallized salts of **4-18**–**4-20** take on a crimson hue in the solid state, their dissolution in MeCN results in a deep teal colour as exemplified by the UV-Vis absorption profiles, of which the wavelength corresponding to the maximum absorption (λ_{max}, Table 4.1) falls within the red region of the visible light spectrum. Coincidently, this absorption band aligns with the lowest energy transition of the spectrum and corresponds to the HOMO-LUMO energy gap. Between the derivatives, a bathochromic shift of λ_{max} is observed with the introduction of a more electronegative atom at the basal R position (i.e., H < Cl < F). Even more so, the influence of the thienyl moieties on the electronic

69

structure is noted between **4-20** and its phenyl counterpart (λ_{max} = 620 nm),[27] resulting in a 35 nm shift in λ_{max}. Such narrowing of the energy gap can be rationalized in terms of the influence of the substituents. In other words, the energy level of the LUMO is lowered as the electronegativity of the halogen atoms increases, while a concomitant raising of the HOMO energy level results from the more electron-donating thienyl moieties; such trends have also been noted within the family of pyridine-bridged thiazyls.[30] Furthermore, the redox behaviour of **4-18–4-20** in MeCN solutions was probed by cyclic voltammetry (CV) and referenced to the ferrocene/ferrocenium redox couple as an internal standard;[31] the results of which are presented in Figure 4.2b and Table 4.1. For each compound, two reversible redox processes were observed, corresponding to a one electron oxidation (0/+) and reduction (-/0) process of the bis-TDA framework in the range of 0.355 to 0.505 V and -0.396 to -0.279 V, respectively. Consistent with the less electropositive core afforded by halogenation at the basal position, individual potentials are shifted anodically. Such redox behaviour is comparable to reported pyridine-bridged thiazyl radicals.[26,30,32–34]

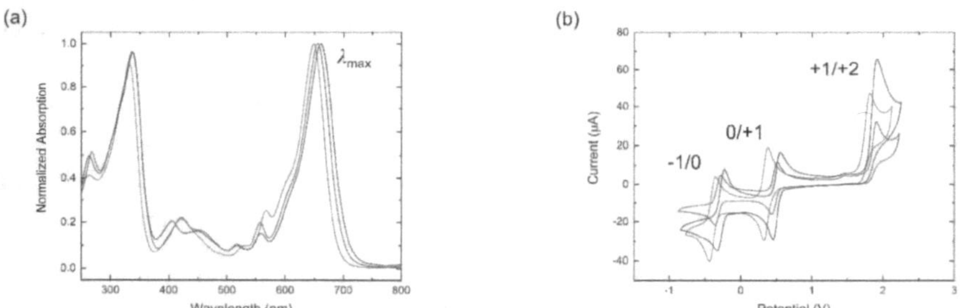

Figure 4.2 (a) UV-Vis spectra and (b) cyclic voltammogram of bis-TDA radicals **4-18** (blue), **4-19** (red) and **4-20** (green) in MeCN.

Table 4.1 Experimental optical[a] and electrochemical[b] parameters of bis-TDA cations **4-18–4-20**.

	4-18 (R = H)	**4-19 (R = F)**	**4-20 (R = Cl)**
λ_{max}	649	661	655
$E_{1/2}^{-1/0}$	-0.396	-0.279	-0.321
$E_{1/2}^{0/+1}$	0.355	0.505	0.475
$E_{1/2}^{+1/+2}{}_c$	1.814	1.916	1.908
$E_{Cell}{}^{d}$	0.750	0.785	0.796

[a] Measured in MeCN and recorded in nm. [b] Measured in MeCN, recorded in V and referenced to the Fc/Fc$^+$ vs. SCE. [b] Irreversible $E_{pc}^{+1/+2}$ reported. [c] $E_{Cell} = E_{1/2}^{0/+1} - E_{1/2}^{-1/0}$.

4.2.2 Reduction and EPR Spectroscopy of bis-TDA Radicals

Based on the 0/+1 redox potential, reduction of **4-18–4-20** was successfully achieved in MeCN using 1,1'-dimethylferrocene (DiMFc) as a reducing agent, affording **4-7–4-9** as maroon microcrystalline solids. To confirm the identity of the radicals, single-crystal X-ray analysis was employed (*vide infra*) along with room temperature X-band EPR spectroscopic studies on dilute solutions of **4-7–4-9** in degassed dichloromethane (DCM). The EPR spectra of each derivative are presented in Figure 4.3a and spectroscopic parameters listed in Table 4.2. Each EPR spectrum consists of a series of complex multiplets with g-values in the range of 2.0031–2.0033. These values slightly deviate from that of a free electron (g_e = 2.0023)[35], reflecting the perturbation from spin-orbit coupling with the heavier sulfur atoms. To extract experimental parameters, the EPR spectra were simulated using a model based on hyperfine coupling to a pair of inequivalent ^{14}N nuclei in each thiadiazinyl moiety (i.e., a_{N1} and a_{N2}). These values are significantly reduced in comparison to bridgeless thiadiazinyl radicals,[36] which is to be expected given the singly occupied molecular orbital (SOMO; see Figure 4.3b) of bis-TDA is partitioned between the two thiadiazinyl rings. Additional couplings were also extracted for spin-active nuclei lying in nodal regions of the SOMO, including the pyridyl nitrogen atom (a_{N3}), basal R substituent (a_R), and the methyl protons ($a_{H(Me)}$), as a result of spin polarization mechanisms and hyperconjugative effects depicted by the spin density plot (Figure 4.3b).[37–39] The weaker hyperfine coupling associated with the chlorine atom, in addition to the quadrupolar nuclear interactions arising from the two abundant isotopes of chlorine (i.e., ^{35}Cl and ^{37}Cl; a_{Cl} = 0.267 G), accounts for the unresolved spectrum exhibited by **4-9**, in comparison to **4-7** and **4-8**. In the case of **4-7**, spin density to the exocyclic thienyl moieties is also detected, contributing to the finer structure of the spectrum from the additional hyperfine interactions with the ^{1}H nuclei ($a_{H(A-C)}$ = 0.216 G and 0.146 G). Overall, these values are in good agreement with the isotropic Fermi contacts calculated through DFT methods on geometry optimized structures of **4-7**–**4-9**, using the unrestricted B3LYP/EPR-II//B3LYP/6-311G(d,p) level of theory.

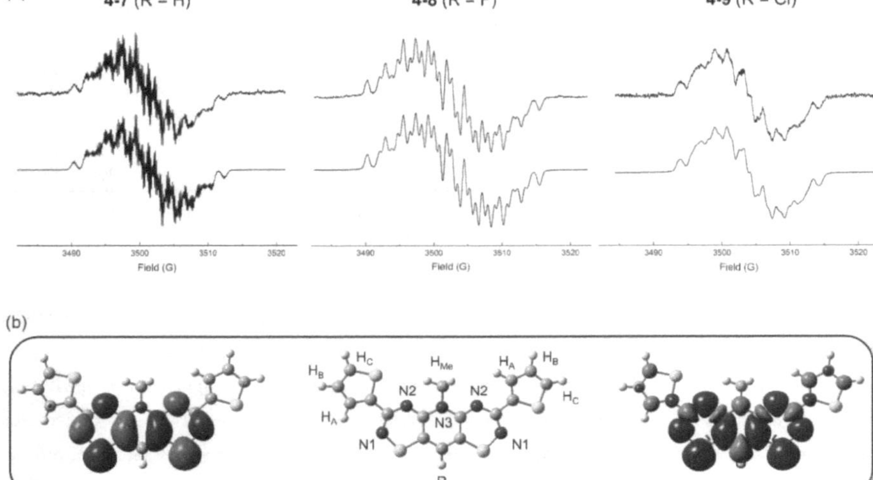

Figure 4.3 (a) Experimental (top) and simulated (bottom) spectra of bis-TDA radicals **4-7** (R = H), **4-8** (R = F) and **4-9** (R = Cl) in DCM. (b) UB3LYP/6-311G(d,p) calculated SOMO (left) and spin density (right) plots of geometry optimized **4-7**. Labelling scheme for coupling constants is presented in the middle.

Table 4.2 EPR parameters^a and calculated isotropic Fermi contacts^b of bis-TDA radicals **4-7–4-9**.

	4-7 (R = H)	**4-8** (R = F)	**4-9** (R = Cl)
a_{N2}	2.686 (2.490, 2.632)	2.673 (2.424, 2.548)	2.696 (2.281, 2.400)
a_{N1}	1.863 (1.896, 2.144)	1.825 (1.824, 2.046)	1.873 (1.730, 1.946)
a_{R}	1.842 (2.669)	5.373 (-6.182)	0.267 (-0.381)
a_{N3}	0.666 (-0.956)	0.623 (-0.816)	0.697 (-0.824)
$a_{H(Me)}$	0.346 (-0.038, -0.920)	0.282 (-0.030, -0.702)	0.306 (-0.031, -0.773)
$a_{H(A,C)}$	0.216 (0.304, 0.400)	-----	-----
$a_{H(B)}$	0.146 (-0.124, -0.146)	-----	-----
g	2.0033	2.0032	2.0031
LW	0.162	0.481	0.510
LS	0.792	0.797	0.797

^{a} Measured in DCM and recorded in units of Gauss. LW = linewidth. LS = line shape. ^{b} UB3LYP calculated coupling constants (in parentheses) using EPR-II basis set for C, H, N, F and 6-311G(d,p) basis set for S, Cl. Geometry optimization performed without symmetry constraints.

4.2.3 Crystal Growth and Phase Separation of bis-TDA Radicals

As mentioned, reduction of bis-TDA cations **4-18–4-20** was successfully achieved in MeCN using DiMFc affording **4-7–4-9** as maroon microcrystalline solids. While chemically pure material could be isolated from bulk reduction reactions, a range of crystallization techniques in MeCN were employed in order to elucidate the structure and confirm the respective identities of **4-7–4-9**, which led to the isolation of a number of polymorphs. For example, slow diffusion of solution of **4-7** into a solution of DiMFc in MeCN at room temperature produced a mixture of needle- and block-like crystals, wherein two crystallographic phases (i.e., α and β, respectively) were determined through single-crystal X-ray diffraction (SCXRD; *vide infra*). Comparison of the powder X-ray diffraction (PXRD) data of the maroon microcrystalline solid isolated from the bulk reduction of **4-7** with the calculated PXRD pattern from the SCXRD data (Figure 4.4a) reveals the bulk reduction reaction exclusively affords the α-phase (α-**4-7**). Carrying out the bulk reduction reaction at elevated temperatures (i.e., 60 °C) selectively forms the β-phase (β-**4-7**), as confirmed through evaluation of the PXRD with that predicted from SCXRD. Further confirmation of phase purity within the bulk samples was facilitated through IR spectroscopy, in which a distinguishable difference exists between the two phases, as provided in Figure 4.5a. Alternatively, recrystallization of **4-7** can be achieved in hot DCE, affording needle-like crystals in which a third crystallographic phase (γ-**4-7**) was identified along with a solvate (**4-7**·DCE). Unfortunately, due to the similarity between the crystal morphology of the γ-**4-7** and **4-7**·DCE, separation was not possible. As well, a fourth phase (δ-**4-7**) was observed in the form of rhombohedral plates upon recrystallization; however, crystals suitable for SCXRD remain elusive.

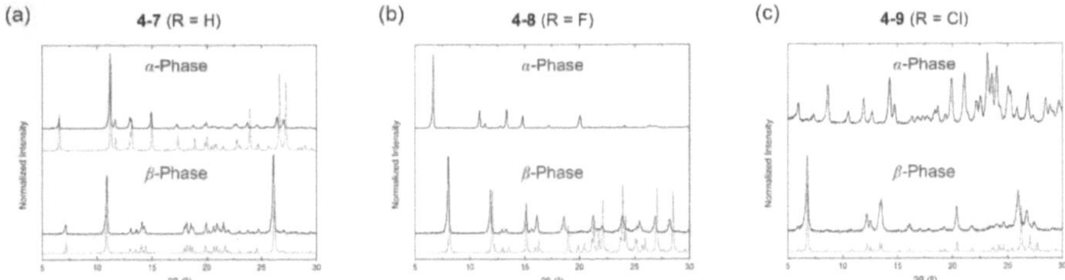

Figure 4.4 PXRD patterns of experimental (solid lines) (a) α- and β- **4-7** (R = H) (b) β-**4-8** (R = F) and (c) β-**4-9** (R = Cl) overlaid with their calculated pattern (dotted lines) from their respective single crystals at 213, 293, 213 and 200 K. Calculated patterns could not be obtained for α-**4-8** and α-**4-9**.

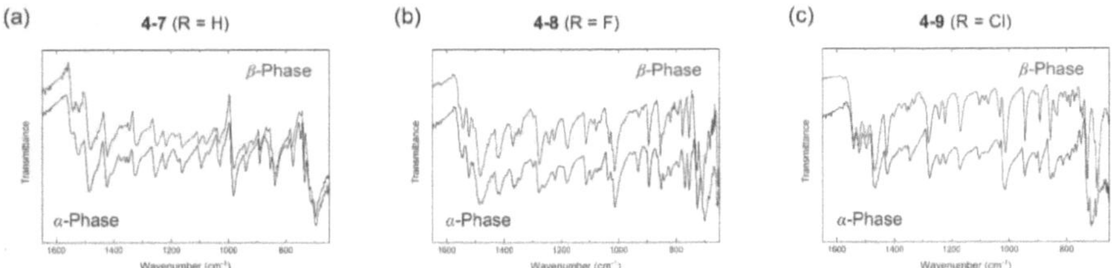

Figure 4.5 FTIR spectra of α- and β- (a) **4-7** (R = H) (b) **4-8** (R = F) and (c) **4-9** (R = Cl).

Aware of the polymorphic nature of **4-7**, reduction reactions of **4-8** and **4-9** with DiMFc were carried out in acetonitrile over a temperature range. From PXRD and FTIR spectroscopic studies (Figures 4.4b,c and 4.5b,c, respectively), bulk material of both analogues collected at 0 °C and at refluxing temperatures inferred the presence of temperature-dependent polymorphs. As such, in pursuit of single crystals suitable for SCXRD, a number of crystallization techniques were employed. For example, standard recrystallization of **4-8** in hot DCE afforded a mixture of needle-like crystals consisting of both the α- and β-phases (α- and β-**4-8**); however, if the temperature of the solution was maintained at 60°C, selective crystallization of β-**4-8** occurred, affording crystals suitable for X-ray analysis. Alternatively, diluting the solution and allowing it to cool to room temperature afforded α-**4-8** selectively. Despite successful isolation of α-**4-8**, crystals of sufficient size for single-crystal X-ray analysis remain elusive. Slow diffusion methods involving solutions of **4-19** and DiMFc were also explored to obtain suitable crystals of α-**4-8** but these proved ineffective towards crystallization due to the limited solubility of **4-8** at room temperature, leading to immediate precipitation of a maroon powder. While similar challenges were encountered with the chloro-derivative **4-9** (i.e., crystals of the α-phase suitable for SCXRD remain elusive), dynamic vacuum sublimation in the temperature range of 100-210°C afforded needle-like crystals of the high temperature β-phase (β-**4-9**).

4.2.4 Structural Analysis of bis-TDA Radicals

Structural analysis of radicals α-, β-, γ-**4-7**, **4-7**·DCE, β-**4-8**, and β-**4-9** was achieved through SCXRD in order to probe the self-assembly of molecules with respect to exocyclic modifications. Crystallographic data are listed in Tables A1.8–A1.9 (see Appendix A1), while pertinent measurements regarding intramolecular and intermolecular parameters are listed in Table 4.3. For each derivative, the asymmetric unit is comprised of a unique discrete radical. Molecularly, the bis-TDA frameworks do not differ all that much from each other as intramolecular bond lengths and angles are consistent amongst the derivatives. Additionally, a slight bowtie-like distortion is adopted along each fused ring system, as viewed down the short molecular axis (Figure 4.6), resulting from an out-of-plane twisting of the sulfur atoms (Figure 4.6). This structural perturbation was measured utilizing the torsion angle ω (Figure 4.6; dashed S–N–N–S connectivity), in which the value of this angle is largest for γ-**4-7** (ω = 5.42°). Although each derivative exhibits a small degree of twisting, as highlighted by ω, the molecules are relatively planar to within 0.052 Å (planarity index, ΔP). Furthermore, the thienyl moieties remain relatively coplanar to the central bis-TDA core with mean torsion angles, denoted by φ in Figure 4.6, within 7°, suggesting the spin density can also be distributed into the thienyl moieties in the solid state.

Table 4.3 Intramolecular and intermolecular parameters[a] for bis-TDA radicals.

	α-**4-7** (R = H)	β-**4-7** (R = H)	γ-**4-7** (R = H)	**4-7**·DCE (R = H)	β-**4-8** (R = F)	β-**4-9** (R = Cl)
ω (°)	-5.0	-2.3	5.4	-2.6	4.0	2.4
ΔP (Å)[b]	0.039	0.021	0.052	0.029	0.040	0.020
φ (°)[c]	-1.3 0.7	3.2 1.9	6.4 1.8	-4.4 0.2	2.0 -1.3	3.7 1.7
τ (°)	62.5	26.0	38.2	24.4	43.2	62.0
d_π (Å)	3.368(3)[d] 3.483(4)[d]	3.390(2)	3.3442(11)	3.467(4)	3.4766(12)	3.4447(11)
d_x (Å)	3.53[e]	6.94	3.89	6.58	3.41	1.51
d_y (Å)	0.54[e]	0.39	1.76	3.52	1.48	1.03

[a] See text for definitions. [b] Planarity index refers to the root mean square deviation of atomic positions to the mean bis-TDA framework. [c] Two values for one unique molecule. [d] Two alternating interplanar distances present. [e] Displacements correspond to radicals separated by the sum of $d_{\pi 1}$ and $d_{\pi 2}$.

Figure 4.6 Two views of γ-**4-7** illustrating (left) the out-of-plane distortion along the bis-TDA framework down the short molecular axis (denoted by the red and green arrows; thienyl substituents removed for clarity) and (right) the torsion angles ω and φ referring to the S–N–N–S connectivity (dashed lines) and mean rotation of the thienyl rings to the core (pink arrow), respectively.

Descriptions of the crystal structures with illustrations are presented below for each phase. While the basis of each solid-state structure consists of slipped π-stacks, regardless of the phase (e.g., α, β, etc.), the degree and alignment of slippage is strongly influenced by the substituents and crystallization technique employed. As such, to lucidly describe the various structures, a number of parameters have been defined (see Figure 4.7). These parameters have been chosen to define the relative position of the bis-TDA core with respect to neighbouring molecules along a π-stack using a local Cartesian coordinate system. The angle in which the mean plane of the bis-TDA core is inclined to the stacking axis is defined by τ and reflects the degree of slippage from the contributions of the interplanar separation (d_π; z-direction) and the displacements along the long (d_x; x-direction) and short (d_y; y-direction) molecular axes. Thus, for a superimposed stack of radicals (i.e., $d_x = d_y = 0$), the heterocyclic bis-TDA planes would be positioned perpendicular to the stacking axis (i.e., $\tau = 90°$).

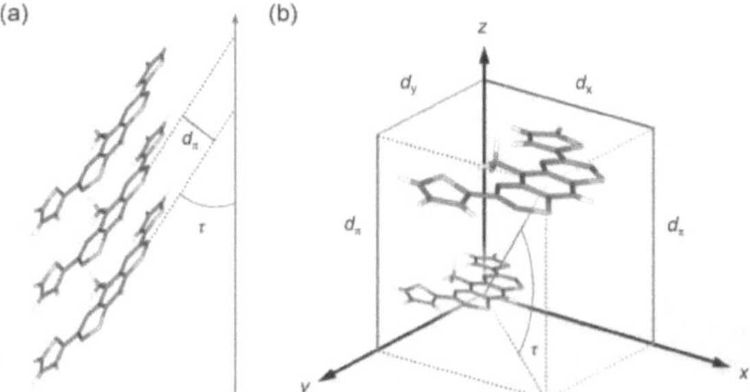

Figure 4.7 Definitions of stacking parameters. (a) One-dimensional array of slipped bis-TDA radicals separated by a plate-to-plate distance denoted by d_π and inclined to the stacking axis (green arrow) by the angle τ. (b) The degree of slippage between bis-TDA radicals relative to a local Cartesian coordinate system with the short and long molecular axes aligning with the x- and y-directions, respectively.

In the low temperature polymorph α-**4**-**7** (R = H), crystals belonging to the orthorhombic space group *Pbcn* comprise grid-like arrays of nonuniform π-stacks running parallel to the a-axis. A bird's-eye view down the stacking direction in Figure 4.8a reveals the curve-shaped molecules overlaying in a twisted head-over-head conformation. From this twisted arrangement, the thienyl moieties can be seen linking molecules of neighbouring stacks in two different manners *via* electrostatic close contacts. Along the z-direction, radicals organize into coplanar arrays like loose-fitting puzzle pieces connected by a weak lateral C–H$\cdots$S' interaction (d_1), nominally larger than the van der Waals (vdW) radii.[40,41] Perpendicular to this, molecules are woven into chain-like arrays along the y-direction in a head-to-tail fashion and tied by short S$\cdots$S' and S$\cdots$N' contacts (d_{2-3}), such that, when viewed down this direction (Figure 4.8b), a cross-braced pattern is generated between adjacent slipped π-stacks. Regarding the nonuniform column itself, stacked molecules are related by a C_2 rotation and separated by two interplanar distances alternating in an ABABAB pattern ($d_{\pi 1}$ = 3.368(3) Å and $d_{\pi 2}$ = 3.483(4) Å). The degree of slippage, depicted by the angle of inclination of τ = 62.5°, imparts two distinct modes of stacking for each radical as shown in Figure 4.8c. The AB pair, which separates radicals by $d_{\pi 1}$, displays a slightly staggered conformation of the π-systems by ~17° with short interannular contacts between bis-TDA frameworks (d_{4-5}). Conversely, the BA radical pair, separated by $d_{\pi 2}$, exhibits a fan-shaped overlay of molecules that concentrates π–π interactions between carbon atoms along the outer regions of the thiadiazinyl and thienyl rings (d_6).

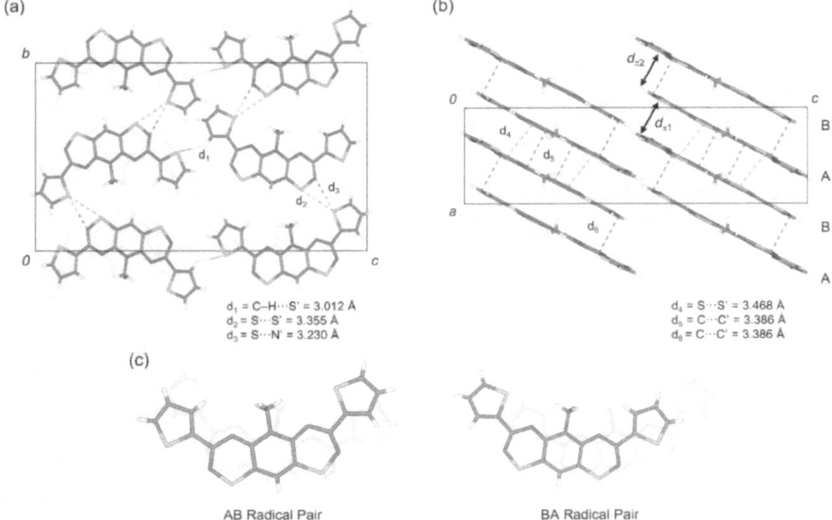

Figure 4.8 Unit cell projections of α-**4-7** illustrating (a) both the coplanar and chain-like arrays linking π-stacked radicals by the respective interstack contacts d_1 and $d_{2,3}$ and (b) the cross-braced packing of the ABABAB slipped π-stacks. (c) Overlay of AB and BA radical pairs along π-stacks. Contacts shown as dash lines and faded regions indicate molecules in background of view. Colour code: C, grey; H, white; N, blue; S, yellow.

By contrast, radicals of the high temperature polymorph β-**4-7** (R = H), crystallizing in the lower-symmetry triclinic space group $P\bar{1}$, are constructed into regular slipped π-stacks that lock within a lamellar arrangement lattice-wide. Here, the plate-to-plate separation between stacked molecules along the x-direction (d_π = 3.390(2) Å) is comparable to the AB radical pair of α-**4-7**, however, a drastic reduction in the stacking inclination is observed by τ = 26.0° rendering less superimposed π-stacks (*cf.* τ = 62.5° for α-**4-7**). From the unit cell projection of the *bc*-plane in Figure 4.9a, the lamellar arrangement of stacks can be seen originating from molecules interlocked laterally into coplanar sheets by a network of C–H···N' and C–H···S' interactions (2.631–3.094 Å). Each sheet consists of molecules aligning lengthwise into unevenly spaced rows parallel to the [02-2] direction that stagger perpendicularly. The discontinuity along a row is reflected in the arrangements of adjacent radicals on alternate sides. In other words, radicals about inversion points at b = ½ (i.e., stacks A and B) are laced closer together by the lateral contacts, while those found about inversion points at c = ½ (i.e., stacks B and A') are further apart, from where no vdW contacts exist. Viewing a row of slipped π-stacks from the side, as shown in Figure 4.9b, illustrates the severe displacement between molecules along the stacking axis (d_x = 6.94 Å; d_y = 0.39 Å), wherein the longitudinal slippage d_x is nearly the length of the bis-TDA core (7.779 Å). Consequentially, this produces a brick-like arrangement of radicals conjoined by a series of π–π contacts (d_{1-5}), of which many comprise of interstack interactions, giving the greater π-coverage as shown in Figure 4.9c.

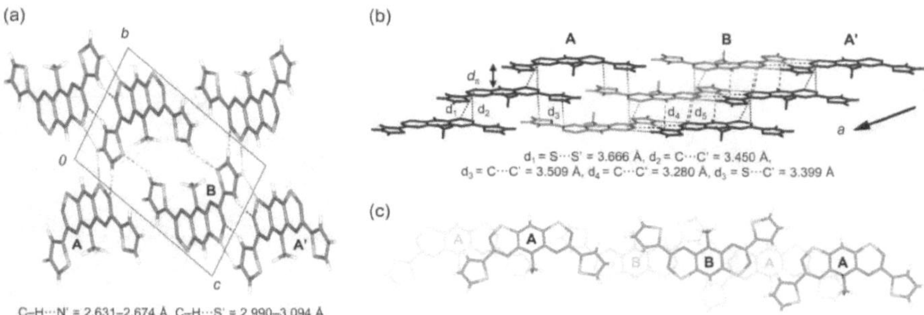

Figure 4.9 (a) Unit cell projection of β-**4-7** illustrating the planar sheets of π-stacked radicals linked by a network of lateral C-H⋯N'/S' contacts. Views of the brick-like arrangement of molecules held together by intra- (d_{1-2}) and interstack (d_{3-5}) π–π contacts (b) from the side (stacks A/A' in dark blue and B in dark yellow for clarity) and (c) down the normal of the bis-TDA plane. Contacts are shown as dash lines and faded regions indicate molecules in background of view. Colour code: C, grey; H, white; N, blue S, yellow.

In the recrystallized polymorph γ-**4-7** possessing the monoclinic space group $P2_1/c$, the assembly of radicals into staggered rows of slipped π-stacks closely resembles that of β-**4-7** from a view down the stacking axis (Figure 4.10a). While the π-stacks have comparable structural features (i.e., d_π = 3.3442(11) Å, τ = 38.2°) to β-**4-7**, adjacent molecules within these rows are related by a 2_1 axis parallel to the b-axis resulting in a herringbone packing pattern (Figure 4.10b). The staggered arrangement of these herringbone arrays along z enables a set of weak lateral S⋯N' contacts (d_1) linking pairs of stacks between rows in a centrosymmetric fashion (e.g., stacks A and B). From an overview of stacks A and B shown in Figure 4.10c, the slippage between radicals (i.e., d_x = 3.89 Å; d_y = 1.76 Å) along a π-stack allows for interstack overlay between sulfur atoms of the bis-TDA frameworks, in which short π-contacts are generated between bis-TDA both along a stack (d_{2-3} = S⋯C') and between stacks (d_4 = S⋯S'). The integrated stacked radicals produce a ladder-like architecture (Figure 4.10d), from which the legs and rungs of the ladder are afforded by the intra- and interstack π–π interactions, respectively.

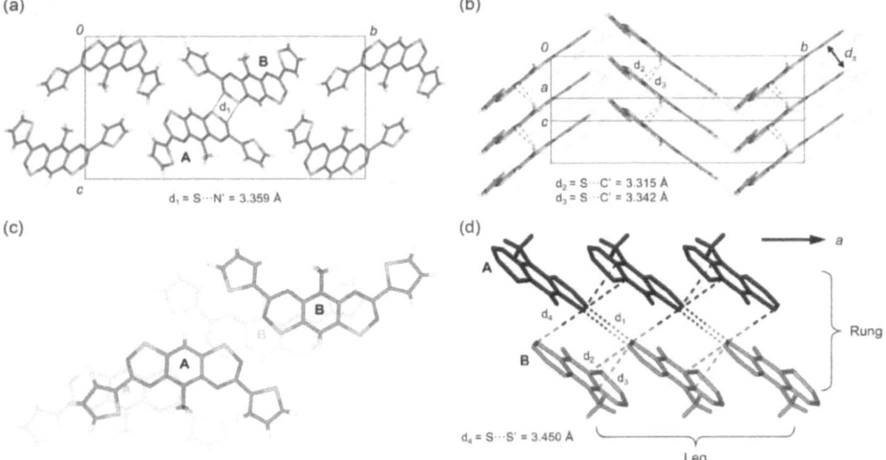

Figure 4.10 Unit cell projections of γ-**4-7** illustrating (a) staggered rows linking stacked molecules centrosymmetrically by contact d_1 and (b) the herringbone packing of slipped π-stacks. Views of the ladder-like arrangement of molecules with intra- (d_{2-3}) and interstack (d_4) π–π contacts (c) down the normal of the bis-TDA plane and (d) sideways (stacks A in dark blue and B in dark yellow; thiophenes omitted for clarity). Contacts are shown as dash lines and faded regions indicate molecules in background of view. Colour code: C, grey; H, white; N, blue; S, yellow.

Ladder-like arrangements are also afforded in the second recrystallized phase **4-7**·DCE by centrosymmetrically linked radicals, of which crystals belong to the monoclinic space group $P2_1/n$. In this case, slipped π-stacks are connected laterally by C–H$\cdots$N' (d_1) across inversion centers at the center and corners of the bc-plane (Figure 4.11a). Along this plane, columns of solvent molecules disrupt the otherwise typical formation of rows along the long molecular axis (e.g., parallel to c). Orthogonally, the radicals of the slipped π-stacks configure into herringbone arrays propagating parallel to the b-axis (Figure 4.11b). By contrast to the previous ladder motifs observed in γ-**4-7**, intrastack radicals are separated by $d_\pi = 3.467(4)$ Å and are inclined by $\tau = 24.4°$, such that, molecular displacement is approximately doubled in each direction (i.e., $d_x = 6.58$ Å; $d_y = 3.52$ Å). As a result, π-overlap is greater between stacks (Figure 4.11c), however, intrastack π–π contacts within the vdW radii are present between carbon atoms of a thienyl ring and the bis-TDA framework (d_2). While these interactions construct the legs of the ladder, centrosymmetric C$\cdots$C' π-contacts (d_{3-5}) comprise the rungs of the ladders between the stacks, some of which are shorter in comparison (Figure 4.11d).

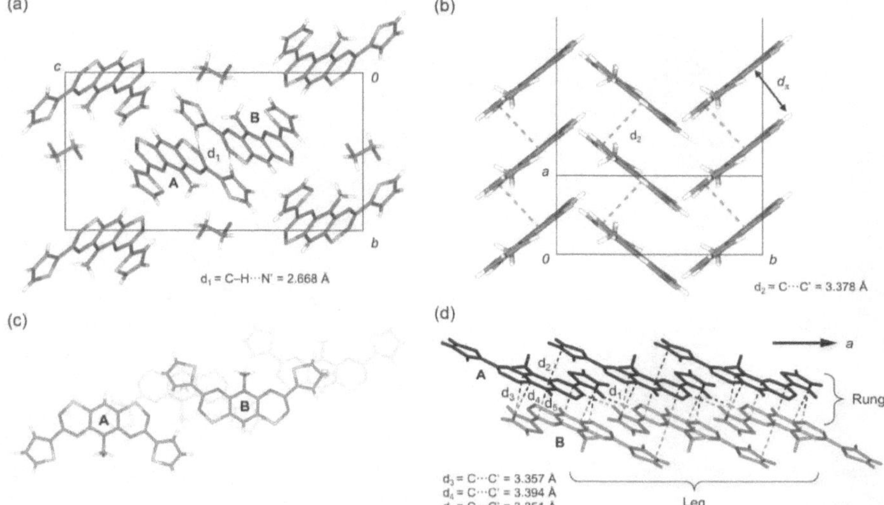

Figure 4.11 Unit cell projections of **4-7**·DCE illustrating (a) stacked molecules linked centrosymmetrically by contact d_1 and (b) the herringbone packing of slipped π-stacks. Views of the ladder-like arrangement of molecules with intra- (d_{2-3}) and interstack (d_4) π–π contacts (c) down the normal of the bis-TDA plane and (d) sideways (stacks A in dark blue and B in dark yellow for clarity). Contacts are shown as dash lines and faded regions indicate molecules in background of view. Colour code: C, grey; H, white; N, blue; S, yellow.

The influences of halogens towards the crystal packing of bis-TDA radicals becomes apparent when analyzing the supramolecular structures. For instance, in crystals of the high temperature phase β-**4-8** (R = F) belonging to the monoclinic space group $P2_1/c$, radicals are assembled evenly into slipped π-stacks ($d_\pi = 3.4766(12)$ Å; $\tau = 43.2°$) along b that are dispersed in grid-like arrays. A top view of the stacks in Figure 4.12a depicts chain-like arrays of neighbouring molecules along c that alternate in a staggered arrangement across the x-direction to produce ribbon-like arrays. Within these chains, radicals are linked by a short contact between the fluoro-substituent and the methyl group (d_1), which leads to an anisotropic head-to-tail alignment. Viewing the crystal structure from a projection of the unit cell from the side, as shown in Figure 4.12b, illustrates the stacked molecules cross-braced down a chain and packed in a herringbone appearance along a ribbon. Additionally, while lateral slippage occurs in both direction along the stacks (i.e., $d_x = 3.41$ Å; $d_y = 1.48$ Å), as highlighted in the overlay of stacked molecules in Figure 4.12c, interstack π–π interactions are not observed and thus, renders isolated one-dimensional π-motifs.

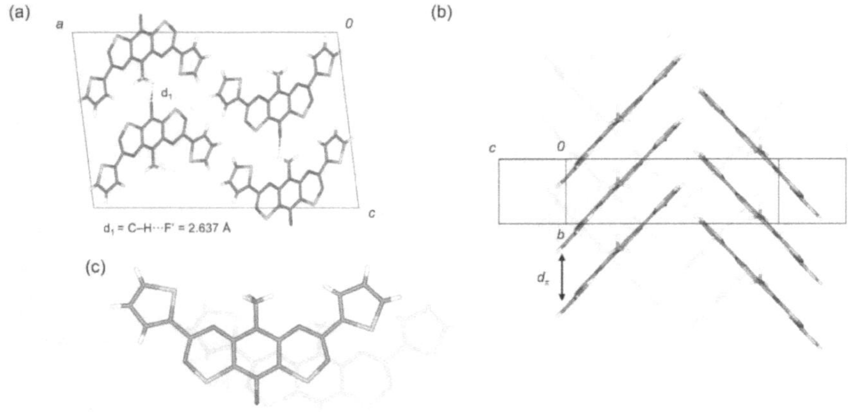

Figure 4.12 (a) Unit cell projection of β-**4-8** illustrating chain-like arrays linking stacked radicals by contact d_1 along c and (b) the herringbone and cross-braced packing of slipped π-stacks. (c) Overlay of adjacent radicals along π-stacks. Contacts are shown as dash lines and faded regions indicate molecules in background of view. Colour code: C, grey; H, white; N, blue; F, pink; S, yellow.

The structural effects from halogen substitution are portrayed in the high temperature phase β-**4-9** (R = Cl) possessing the monoclinic space group $P2_1/n$. Molecules align into slipped π-stacked arrangements along a, which are separated uniformly by $d_\pi = 3.4447(11)$ Å and inclined steeper to the stacking axis by τ = 61.2°. Chains continue to pursue along the z-direction with molecules woven in and out of alignment along the z-direction and connected laterally by short S···S' interactions (d_1) between the thienyl and bis-TDA ring systems (Figure 4.13a). A slight ruffling between neighbouring stacks is observed along the y-direction (Figure 4.13b), in contrast to the more pronounce herringbone arrangement exhibited in previous structures. Furthermore, chains cross-link to produce 2D lateral networks afforded by centrosymmetric S···N' and S···S' contacts ($d_{2–3}$). The interlocking of the centrosymmetric stacks, in accordance with the degree of slippage between stacked radical (Figure 4.13c; $d_x = 1.51$ Å; $d_y = 1.03$ Å), results in a weakly associated ladder-like motif with legs comprised of the intrastacked radicals connected by a S···C' contact (d_4) and rungs created by a second set of intercolumnar S···N' interactions (d_5) outside the vdW radii (Figure 4.13d).

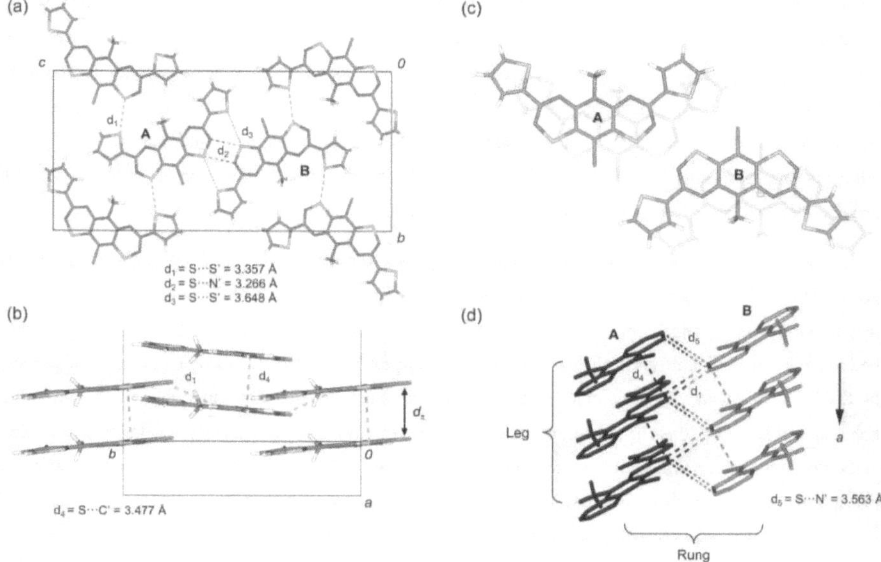

Figure 4.13 Unit cell projections of β-**4-9** illustrating (a) π-stacked radicals woven into chain-like arrays linked by contact d_1 and further cross-linked in two dimensions by contacts $d_{2,3}$ and (b) the ruffling packing of the slipped π-stacks. Views of the ladder-like arrangement of molecules with intra- (d_4) and interstack (d_5) π–π contacts (c) down the normal of the bis-TDA plane and (d) from the side. Contacts shown as dash lines and faded regions indicate molecules in background of view.

From the described crystal structures, a few observations are noted. In all phases, radicals prefer to assemble into slipped π-stacks, which presumably is driven by the apparent steric interactions between N-methyl groups. Polymorphs acquired at elevated temperatures, by either chemical reduction or recrystallization, all display regular columns with equidistant plate-to-plate separations, while a Peierls distortion appears to stabilize the π-stacks of the low temperature phase of the proto- analogue. Of the uniform arrays, stacks belonging to the halogenated derivatives exhibit longer interplanar separations in comparison to those of the single-component phases belonging to the proto- analogue by approximately ~0.1 Å. This is speculated to originate from the lone pair repulsion between halogens at the basal carbon. On the other hand, the degree of slippage along a stack varies between structures and appears to be influenced by several structural factors. Most notable is the curve-shaped extension of the π-surface through the nearly coplanar thienyl rings capable of facilitating lateral displacements in two dimensions (i.e., d_x and

d_y). In congruence, π-stacks in the present structures have a proclivity to pair in a centrosymmetric fashion, in which the irregular π-distribution can perpetuate variances in slippage through interstack π–π contacts. For example, d_x is largest when the bis-TDA frameworks are aligned lengthwise, as seen in β-**4-7** and **4-7**·DCE, where effective π-overlay is established between the two stacks. The remaining phases see π-stacks of this motif pivot away from this arrangement. In doing so, interstack interactions require more contributions from latitudinal slippage (d_y). In contrast to all this, bis-DTA analogues (**4-2**) without the peripheral substituents have, for the most part, demonstrated slippage longitudinally (d_x).[33,42,43]

Additionally, the thienyl moieties impact the supramolecular packing significantly, given its possession of both electron and hydrogen donating sources. Radicals have been shown to lock into herringbone-type arrangements by favourable C–H···π interactions afforded between the exterior heterocycles, although, in some phases, these moieties appear to stabilize anisotropic planar arrangements by C–H···S'/N' contacts. With a plethora of competing interactions, this may rationalize the polymorphic behaviour observed in these systems, specifically **1-7**. Furthermore, the impact of the halogen at the basal carbon on the solid-state assembly seems to have opposing effects. In β-**4-8** (R = F), the crystal maintains the alignment of molecules seemingly from a central-laying charge dipole. On the contrary, molecules of β-**4-8** (R = F) forego less directional order, most likely a response to the size of the chlorine atoms

4.3 Summary and Conclusion

From a material perspective, pyridine-bridged bisthiadiazinyls **4-4** are an appealing class of stable radicals. The basis of the molecular framework possesses three points of functionalization that enable fine-tuning of the structure-property relationships. Among these, the beltline substituents (R^1 and R^2) lie along a nodal plane of the SOMO, from which the effects originating from these groups are significantly influential towards the crystal packing. On the other hand, the peripheral ligand (R^3), directly attached to the spin-bearing framework, and further provides opportunities to modify the electronic character. As a continuation of the work encompassing heteroaromatic functionalization, a series of thienyl-substituted bisthiadiazinyls was reported herein was conducive of that. Distribution of spin density along the thienyl rings was reflected by the rich hyperfine splitting pattern observed in the EPR spectrum of **4-7**, while fine-tuning of the molecular orbital energies *via* halogenation was observed by CV and UV-Vis spectroscopy. Noteworthy, the impact of functionalization towards the solid-state assembly was observed when five crystallographic phases (α, β, γ, δ and DCE solvate) were obtained for the proto- derivative. Among the structures described, the thienyl groups ability to direct crystal packing in several structural architectures (e.g., uniform and non-uniform π-stacks, herringbone and lamellar arrays, brick- and ladder-like arrangements) comes through in its flexibility to engage in favourable π–π and electrostatic interactions. Furthermore, halogenation at the basal carbon (R = F, Cl) can modify the spatial alignments of molecules with respect to the hydrogenated derivative, due to differences in contrasting size and electronegativity of the atoms. Further studies on the functionalization of bisthiadiazinyls are expected to shed new light on novel structural assemblies for thiazyl radicals, which hold promise for the design of molecular conductors, magnets and bistable switches.

4.4 Experimental Details

4.4.1 General Procedures

The reagents benzyltriethylammonium chloride (Alfa Aesar), sulfur monochloride (Sigma-Aldrich) and 1,1'-dimethylferrocene (Alfa Aesar) were obtained commercially and used as received. The remaining starting materials 2-thiophenecarboxamidine,[44] 2,4,6-trifluoropyridine (**4-11**),[45] *N*-methyl-2,6-difluoropyridinium triflate (**4-12**)[30] and *N*-methyl-2,4,6-trifluoropyridinium triflate (**4-13**)[46] were prepared as previously described (Scheme 4.1). All solvents were of reagent grade; MeCN and DCE were dried by passing through activated alumina on a J. C. Meyer solvent purification system. Unless otherwise indicated, all reactions were performed under an atmosphere of dry nitrogen using dried solvents. Melting points were taken using a Mel-Temp apparatus and are uncorrected. NMR (^{1}H, ^{13}C, ^{19}F) spectroscopy was performed on samples dissolved in deuterated solvents at room temperature using a Bruker Avance 400 MHz spectrometer. ^{1}H and ^{13}C NMR spectra were referenced to the residual solvent signal.[47] ^{19}F NMR spectra were referenced to a trifluoroacetic acid internal standard.[48] IR spectroscopy was performed on solid samples using an Agilent Technologies Cary 630 FTIR spectrometer. Elemental analyses were performed by MHW Laboratories, Phoenix, AZ 85018 and Atlantic Microlab, Norcross, GA 30071.

4.4.2 Experimental Procedures

Preparation of 2,6-bis(amino(thiophen-2-yl)methylidene)amino-1-methylpyridin-1-ium trifluoromethanesulfonate (4-14) and 2-(1-(6-(amino(thiophen-2-yl)methylidene)amino)-1-methylpyridin-2(1*H*)-ylidene)amino(thiophen-2-yl)methylidene)amino-6-(amino(thiophen-2-yl)methylidene)amino-1-methylpyridin-1-ium chloride (4-17). A solution of **4-10** (5.40 g, 19.3 mmol) in MeCN (200 mL) was added dropwise to a solution of 2-thiophenecarboxamidine (10.0 g, 79.3 mmol) in MeCN (200 mL). After the addition, the resulting yellow slurry was filtered and washed with DCM. The filtrate was dried *in vacuo* resulting in an orange solid mixture of **4-14** and **4-16**. DCM was added to the crude solid forming a yellow slurry that was cooled to -20°C for 1 h. The slurry was filtered cold to remove excess **4-16**, washed with cold DCM and dried. Recrystallization of the crude filtered product from water afforded yellow needle-like crystals of **4-14**. Yield 5.13 g (10.4 mmol; 54%). MP = 188–189 °C. ^{1}H NMR (δ, CD$_3$CN): 7.94 (1H, t, *J* = 8.2 Hz), 7.75 (2H, dd, *J* = 3.8, 1.1 Hz), 7.71 (2H, dd, *J* = 5.1, 1.1 Hz), 7.20 (2H, dd, *J* = 5.1, 3.8 Hz), 6.94 (2H, d, *J* = 8.2 Hz), 6.60 (4H, s). 3.77 (3H, s). ^{13}C NMR (δ, CD$_3$CN): 157.54 (2C) 155.16 (2C), 145.16 (1CH), 139.12 (2C), 133.32 (2CH), 130.42 (2CH), 129.32 (2CH), 122.09 (1C, q,

J_{CF} = 316.3 Hz), 112.10 (2CH), 35.22 (1CH$_3$). ^{19}F NMR (δ, CD$_3$CN): -79.10 (3F, s). IR (ν_{max}, ATR): 3407 (m), 3337 (m), 3226 (m), 3088 (w), 2954 (w), 1657 (m), 1649 (m), 1598 (s), 1531 (s), 1476 (m), 1433 (s), 1400 (m), 1365 (w), 1356 (w), 1273 (m), 1258 (s), 1244 (s), 1232 (s), 1224 (s), 1187 (w), 1168 (s), 1144 (s), 1114 (m), 1080 (w), 1045 (w), 1028 (s), 920 (m), 859 (w), 850 (m), 828 (m), 814 (w), 775 (w), 762 (m), 753 (w), 708 (s), 671 (w) cm^{-1}. Anal. Calcd for C$_{17}$H$_{16}$F$_3$N$_5$O$_3$S$_3$: C, 41.54; H, 3.28; N, 14.25%. Found: C, 41.59; H, 3.23; N, 14.21%. To confirm the identity of **4-16**, DCM from the filtrate of the step above was eliminated under reduced pressure. The crude solid was dissolved in MeCN, to which excess triethylbenzylammonium chloride was added and stirred for 30 minutes. The resulting orange precipitate was filtered, washed with MeCN and dried. Recrystallization of the crude solid from isopropanol afforded yellow needles of **4-17**. ^{1}H NMR (δ, (CD$_3$)$_2$SO): 8.00 (2H, dd, J_{HH} = 5.0, 1.1 Hz), 7.90 (4H, s), 7.77 (1H, dd, J_{HH} = 5.0, 1.2 Hz), 7.64 (1H, dd, J = 3.7, 1.2 Hz), 7.58 (2H, dd, J = 8.5, 7.9 Hz), 7.58 (2H, dd, J = 5.0, 3.8 Hz), 7.14 (1H, dd, J = 3.8, 1.2 Hz), 6.60 (2H, dd, J = 8.5, 1.0 Hz), 6.43 (2H, dd, J = 7.9, 1.0 Hz), 3.79 (6H, s). ^{13}C NMR (δ, (CD$_3$)$_2$SO): 157.57 (1C), 156.48 (2C), 154.74 (2C), 153.43 (2C), 143.43 (1C), 141.33 (2CH), 138.90 (2C), 131.98 (2CH), 131.48 (1CH), 130.42 (1CH), 129.57 (2CH), 128.21 (2CH), 128.08 (1CH), 108.43 (2CH), 105.90 (2CH), 33.46 (2CH$_3$).

Preparation of 2,6-bis(amino(thiophen-2-yl)methylidene)amino-4-fluoro-1-methylpyridin-1-ium trifluoromethanesulfonate (4-15). Prepared analogously to **4-14** starting from **4-11** (5.73 g, 19.3 mmol) to afforded faint yellow shard-like crystals of **4-15**. Yield 4.12 g (8.09 mmol; 42%). MP = 175–177 °C. ^{1}H NMR (δ, CD$_3$CN): 7.78 (2H, dd, J_{HH} = 3.8, 1.1 Hz), 7.74 (2H, dd, J_{HH} = 5.1, 1.1 Hz), 7.21 (2H, dd, J_{HH} = 5.1, 3.8 Hz), 6.78 (4H, s), 6.74 (2H, d, J_{HF} = 8.6 Hz), 3.72 (3H, s). ^{13}C NMR (δ, CD$_3$CN): 173.05 (1C, d, J_{CF} = 266.3 Hz), 159.89 (2C, d, J_{CF} = 16.1 Hz), 155.79 (2C), 138.80 (2C), 133.72 (2CH), 130.79 (2CH), 129.43 (2CH), 122.02 (1C, q, J_{CF} = 318.2 Hz), 99.66 (2CH, d, J_{CF} = 24.1 Hz), 35.00 (1CH$_3$, s). ^{19}F NMR (δ, CD$_3$CN): -79.12 (3F, s), -92.32 (1F, t, J_{FH} = 8.6 Hz). IR (ν_{max}, ATR): 3408 (m), 3340 (m), 3227 (m), 3100 (m), 1648 (m), 1604 (m), 1572 (m), 1534 (m), 1509 (m), 1475 (m), 1428 (s), 1401 (s), 1355 (s), 1309 (m), 1269 (s), 1247 (s), 1215 (s), 1197 (s), 1163 (s), 1141 (s), 1041 (m), 1028 (s), 1012 (m), 964 (m), 836 (m), 826 (m), 755 (w), 707 (s) cm^{-1}. Anal. Calcd for C$_{17}$H$_{15}$F$_4$N$_5$O$_3$S$_3$: C, 40.07; H, 2.97; N, 14.91%. Found: C, 39.06; H, 2.85; N, 12.94%.

Preparation of 5-methyl-3,7-di(thiophen-2-yl)-[1,2,4]thiadiazino[6',5':5,6]pyrido[2,3-e][1,2,4]thiadiazin-1-ium trifluoromethanesulfonate (4-18). Sulfur monochloride (2.20 g, 16.3 mmol) was added to a stirring slurry of **4-14** (2.00 g, 4.07 mmol) in chlorobenzene (100 mL) and refluxed gently for 2 h. The resulting red microcrystalline precipitate was filtered hot, washed with DCE and dried *in vacuo*. Recrystallization of the crude product from MeCN afforded ruby-coloured crystals of **4-18**. Yield 0.81 g (1.47 mmol; 36%). dec. > 242 °C. ^{1}H NMR (δ, CD$_3$CN): 8.00 (2H, dd, J = 3.8, 1.2 Hz), 7.84 (2H, dd, J =

5.0, 1.2 Hz), 7.26 (2H, d, J = 5.0, 3.8 Hz), 3.80 (3H, s). IR (ν_{max}, ATR): 3078 (w), 1520 (m), 1484 (s), 1421 (m), 1393 (m), 1361 (s), 1320 (m), 1257 (s), 1221 (m), 1208 (m), 1149 (s), 1089 (m), 1038 (m), 1027 (m), 993 (s), 943 (m), 878 (w), 856 (m), 839 (m), 806 (w), 749 (s), 731 (s), 702 (s) cm^{-1}. Anal. Calcd for $C_{17}H_{10}F_3N_5O_3S_5$: C, 37.15; H, 1.83; N, 12.74 %. Found: C, 37.03; H, 1.69; N, 12.83%.

Preparation of 10-fluoro-5-methyl-3,7-di(thiophen-2-yl)-[1,2,4]-thiadiazino[6',5':5,6]pyrido[2,3-e][1,2,4]thiadiazin-1-ium trifluoromethanesulfonate (4-19). Sulfur monochloride (0.53 g, 3.9 mmol) was added to a slurry of **4-15** (0.50 g, 0.98 mmol) in a mixed MeCN and DCE solvent system (1:19, 20 mL) and refluxed gently for 2 h. The resulting red microcrystalline precipitate was filtered cold, washed with DCE and carbon disulfide, and dried *in vacuo*. Recrystallization of the crude product from MeCN afforded ruby-coloured crystals of **4-19**. Yield 0.14 g (0.25 mmol; 25%). dec. < 286 °C. ^{1}H NMR (δ, CD$_3$CN): 8.06 (2H, dd, J = 3.8, 1.2 Hz), 7.90 (2H, dd, J = 5.0, 1.2 Hz), 7.23 (2H, d, J = 5.0, 3.8 Hz), 3.84 (3H, s). IR (ν_{max}, ATR): = 3085 (w), 1567 (w), 1523 (m), 1479 (m), 1458 (m), 1411 (m), 1392 (m), 1365 (m), 1355 (m), 1328 (m), 1273 (m), 1250 (m), 1206 (m), 1150 (m), 1126 (w), 1086 (w), 1071 (w), 1028 (s), 953 (w), 936 (w), 883 (w), 655 (m), 836 (m), 803 (w), 768 (m), 742 (m), 722 (s), 702 (s), 661 (s) cm^{-1}. Anal. Calcd for $C_{17}H_9F_4N_5O_3S_5$: C, 35.97; H, 1.60; N, 12.34 %. Found: C, 35.86; H, 1.43; N, 12.40%.

Preparation of 10-chloro-5-methyl-3,7-di(thiophen-2-yl)-[1,2,4]-thiadiazino[6',5':5,6]pyrido[2,3-e][1,2,4]thiadiazin-1-ium trifluoromethanesulfonate (4-20). Sulfur monochloride (2.20 g, 16.3 mmol) was added to a slurry of **4-14** (2.00 g, 4.07 mmol) in dry MeCN (15 mL) and refluxed gently for 16 h. The reaction was cooled to 0°C for 30 minutes and the resulting red microcrystalline precipitate was filtered cold, washed with DCE and carbon disulfide, and dried *in vacuo*. Recrystallization of the crude product from MeCN afforded ruby-coloured crystals of **4-19**. Yield 0.85 g (1.45 mmol; 36%). dec. > 281°C. ^{1}H NMR (δ, CD$_3$CN): 8.06 (2H, dd, J = 3.8, 1.2 Hz), 7.90 (2H, dd, J = 5.0, 1.2 Hz), 7.23 (2H, d, J = 5.0, 3.8 Hz), 3.84 (3H, s). (ν_{max}, ATR): 3090 (w), 1522 (m), 1499 (m), 1471 (s), 1422 (m), 1395 (m), 1360 (s), 1325 (m), 1260 (s), 1227 (s), 1210 (m), 1161 (m), 1140 (s), 1120 (m), 1092 (m), 1027 (s), 963 (m), 856 (m), 802 (w), 760 (m), 748 (s), 741 (s), 727 (s), 705 (s), 658 (m) cm^{-1}. Anal. Calcd for $C_{17}H_9ClF_3N_5O_3S_5$: C, 34.96; H, 1.55; N, 11.99 %. Found: C, 34.98; H, 1.54; N, 12.11%.

Preparation of α-5-methyl-3,7-di(thiophen-2-yl)-[1,2,4]thiadiazino[6',5':5,6]pyrido[2,3-e][1,2,4]thiadiazin-2-yl (α-4-7). *Method 1: Bulk material.* DiMFc (0.137 g, 0.64 mmol) was added to a stirring, nitrogen-sparged solution of **4-18** (0.25 g, 0.45 mmol) in MeCN (60 mL) at 0 °C. After 1 h, the resulting maroon microcrystalline precipitate was filtered, washed with MeCN and dried *in vacuo*. Yield 0.13 g (0.32 mmol; 71%). IR: See Figure 4.5a for spectrum. Anal. Calcd for $C_{16}H_{10}N_5S_4$: C, 47.98; H, 2.52; N, 17.49%. Found: C, 47.68; H, 2.39; N, 17.29%. *Method 2: Single crystals for X-ray work.* Ice-cold, degassed solutions (three freeze-pump-thaw cycles) of DiMFc (14.6 mg, 0.07 mmol) in MeCN (10 mL) and **4-18** (25.0 mg, 0.05 mmol) in MeCN (20 mL) were allowed to diffuse together slowly over a period of 48 h. The solvent was decanted to leave golden needles of α-**4-7**.

Preparation of β-5-methyl-3,7-di(thiophen-2-yl)-[1,2,4]thiadiazino[6',5':5,6]pyrido[2,3-e][1,2,4]thiadiazin-2-yl (β-4-7). *Method 1: Bulk material.* Prepared analogously to α-**4-7** at 60 °C resulting in a maroon microcrystalline precipitate of β-**4-7**. Yield 0.11 g (0.27 mmol; 58%). IR: See Figure 4.5a for spectrum. *Method 2: Single crystals for X-ray work.* Degassed solutions (three freeze-pump-thaw cycles) of DiMFc (11.7 mg, 0.11 mmol) in MeCN (10 mL) and **4-18** (40.0 mg, 0.07 mmol) in MeCN (20 mL) were allowed to diffuse together slowly over a period of 16 h. The solvent was decanted to leave golden needles of β-**4-7**.

Preparation of β-10-fluoro-5-methyl-3,7-di(thiophen-2-yl)-[1,2,4]thiadiazino[6',5':5,6]pyrido[2,3-e][1,2,4]thiadiazin-2-yl (β-4-8). Prepared analogously to β-**4-7** starting with **4-19** (0.25 g, 0.44 mmol) and DiMFc (0.141 g, 0.66 mmol) below reflux resulting in a maroon precipitate of β-**4-8**. Yield 0.18 g (0.43 mmol; 98%). Single crystals for X-ray work were grown by slow recrystallization from refluxing DCE cooling to 60°C. IR: See Figure 4.5b for spectrum. Anal. Calcd for $C_{16}H_9FN_5S_4$: C, 45.92; H, 2.17; N, 16.73%. Found: C, 41.68; H, 1.63; N, 15.19%.

Preparation of β-10-chloro-5-methyl-3,7-di(thiophen-2-yl)-[1,2,4]thiadiazino[6',5':5,6]pyrido[2,3-e][1,2,4]thiadiazin-2-yl (β-4-9). Prepared analogously to β-**4-7** starting with **4-20** (0.25 g, 0.43 mmol) and DiMFc (0.137 g, 0.64 mmol) at 60 °C resulting in a maroon precipitate of β-**4-9**. Yield 0.18 g (0.41 mmol; 95%). Single crystals for X-ray work were grown by dynamic vacuum sublimation at 10^{-5} Torr down a temperature gradient of 210–100°C. IR: See Figure 4.5c for spectrum. Anal. Calcd for $C_{16}H_9ClN_5S_4$: C, 44.18; H, 2.09; N, 16.10%. Found: C, 44.05; H, 1.93; N, 15.85%.

4.4.3 UV-Vis Spectroscopy

UV-visible spectra were recorded on samples of dissolved in MeCN at ambient temperature using an Agilent Cary 5000 UV-Vis-NIR spectrophotometer. Measurements were completed in the range 250–800 nm with solutions occupying a standard 10 mm pathlength cuvettes.

4.4.4 Electrochemistry

Cyclic voltammetry was performed using a BASi Epsilon potentiostat employing a glass cell and platinum wires for working, counter and pseudo-reference electrodes. The measurements were carried out on samples dissolved in degassed acetonitrile containing 0.1 M tetrabutylammonium hexafluorophosphate (Oakwood) as supporting electrolyte with a scan rate of 200 mV/s. The experiments were referenced to the Fc/Fc^+ redox couple of ferrocene at 0.382 V vs. SCE.[31]

4.4.5 EPR Spectroscopy

X-band EPR spectra were recorded at ambient temperature using a Bruker EMXplus spectrometer on samples dissolved in degassed dichloromethane. Hyperfine coupling constants were obtained by spectral simulation using the SpinFit module of the Bruker Xenon software.

4.4.6 Powder X-Ray Diffraction

PXRD measurements were performed on bulk powder samples of α-, β-**4-7**, β-**4-8** and β-**4-9** using a Rigaku Ultima IV powder diffractometer with an X-ray source of Cu K_a ($\lambda = 1.5418$ Å) and a θ–2θ geometry. Phases purity of bulk samples was assessed by comparison of PXRD data with the patterns calculated from single-crystal data using the Mercury software program.[49]

4.4.7 Single-Crystal X-Ray Diffraction

Data collection for single crystals of α-, β-, γ-**4-7**, **4-7**·DCE, β-**4-8**, β-**4-9**, **4-17** and **4-19** were collected on a Bruker Smart or Kappa APEX II single crystal diffractometer equipped with a graphite monochromator. Samples were mounted on MiTeGen loop using parabar oil. Both instruments were equipped with a sealed tube Mo K_a source ($\lambda = 0.71073$ Å), an APEX II CCD detector, and a dry compressed air-cooling system. All samples were cooled to 200(2) K during data collection except for β-**4-7**, which remained at room temperature. Raw data collection and processing were performed with the APEX3 software package from Bruker.[50] Semi-empirical absorption corrections based on equivalent reflections were applied.[51] Systematic absences in the diffraction data set and unit-cell parameters were consistent with the assigned space group. The initial structural solutions were determined using SHELXT direct methods[52] and refined with full-matrix least-squares procedures based on F2 using SHELXL or ShelXle.[53] Hydrogen atoms were placed geometrically and refined using a riding model.

4.4.8 Theoretical Calculations

All calculations were performed at the DFT level using the UB3LYP functional, as contained in the Gaussian 09W suite of programs.[54] Molecular geometries were optimized using the 6-311+G(2d,p) basis set without symmetry constraints. Isotropic Fermi contacts were taken from single-point calculations using the double-zeta basis set EPR-II for first-row atoms and split-valence triple-basis set 6-311G(d,p) for second-row atoms on the optimized geometries.

4.5 References

1   R. G. Hicks, *Stable Radicals*, John Wiley & Sons, Ltd, Chichester, UK, 2010.

2   I. Ratera and J. Veciana, *Chem. Soc. Rev.*, 2012, **41**, 303–349.

3   R. C. Haddon, *Nature*, 1975, **256**, 394–396.

4   M. Vérot, J. B. Rota, M. Kepenekian, B. Le Guennic and V. Robert, *Phys. Chem. Chem. Phys.*, 2011, **13**, 6657–6661.

5   A. Caneschi, D. Gatteschi, R. Sessoli and P. Rey, *Acc. Chem. Res.*, 1989, **22**, 392–398.

6   J. S. Miller, *Adv. Mater.*, 2002, **14**, 1105–1110.

7   T. Sugawara and M. M. Matsushita, *J. Mater. Chem.*, 2009, **19**, 1738.

8   Y. Morita, S. Suzuki, K. Sato and T. Takui, *Nat. Chem.*, 2011, **3**, 197–204.

9   J. M. Rawson, A. Alberola and A. Whalley, *J. Mater. Chem.*, 2006, **16**, 2560.

10   R. G. Hicks, *Nat. Chem.*, 2011, **3**, 189–191.

11   B. Tang, J. Zhao, J. F. Xu and X. Zhang, *Chem. Sci.*, 2020, **11**, 1192–1204.

12   K. Kato and A. Osuka, *Angew. Chemie*, 2019, **131**, 9074–9082.

13   D. A. Haynes, *CrystEngComm*, 2011, **13**, 4793–4805.

14   A. W. Cordes, R. C. Haddon and R. T. Oakley, *Phosphorus. Sulfur. Silicon Relat. Elem.*, 2004, **179**, 673–684.

15   A. Mailman, C. M. Robertson, S. M. Winter, P. A. Dube and R. T. Oakley, *Inorg. Chem.*, 2019, **58**, 6495–6506.

16   J. M. Rawson and C. P. Constantinides, in *Materials and Energy*, ed. J. S. Miller, World Scientific Publishing Co. Pte Ltd, 2018, vol. 4, pp. 95–124.

17   K. E. Preuss, *Coord. Chem. Rev.*, 2015, **289–290**, 49–61.

18   K. E. Preuss, *J. Chem. Soc. Dalt. Trans.*, 2007, 2357–2369.

19   D. Tian, S. M. Winter, A. Mailman, J. W. L. Wong, W. Yong, H. Yamaguchi, Y. Jia, J. S. Tse, S. Desgreniers, R. A. Secco, S. R. Julian, C. Jin, M. Mito, Y. Ohishi and R. T. Oakley, *J. Am. Chem. Soc.*, 2015, **137**, 14136–14148.

20   R. G. Hicks, *Org. Biomol. Chem.*, 2007, **5**, 1321–1338.

21   M. A. Nascimento and J. M. Rawson, in *Encyclopedia of Inorganic and Bioinorganic Chemistry*, John Wiley & Sons, Ltd., 2019, pp. 1–11.

22   N. J. Yutronkie, D. Bates, P. A. Dube, S. M. Winter, C. M. Robertson, J. L. Brusso and R.

T. Oakley, *Inorg. Chem.*, 2019, **58**, 419–427.

23   P. J. Hayes, R. T. Oakley, A. W. Cordes and W. T. Pennington, *J. Am. Chem. Soc.*, 1985, **107**, 1346–1351.

24   R. T. Boeré, A. W. Cordes, P. J. Hayes, R. T. Oakley, R. W. Reed and W. T. Pennington, *Inorg. Chem.*, 1986, **25**, 2445–2450.

25   R. T. Boeré, T. L. Roemmele and X. Yu, *Inorg. Chem.*, 2011, **50**, 5123–5136.

26   L. Beer, R. C. Haddon, M. E. Itkis, A. A. Leitch, R. T. Oakley, R. W. Reed, J. F. Richardson and D. G. VanderVeer, *Chem. Commun.*, 2005, 1218–1220.

27   A. A. Leitch, R. T. Oakley, R. W. Reed and L. K. Thompson, *Inorg. Chem.*, 2007, **46**, 6261–6270.

28   A. A. Leitch, I. Korobkov, A. Assoud and J. L. Brusso, *Chem. Commun.*, 2014, **50**, 4934–4936.

29   N. J. Yutronkie, A. A. Leitch, I. Korobkov and J. L. Brusso, *Cryst. Growth Des.*, 2015, **15**, 2524–2532.

30   S. M. Winter, A. R. Balo, R. J. Roberts, K. Lekin, A. Assoud, P. A. Dube and R. T. Oakley, *Chem. Commun.*, 2013, **49**, 1603–1605.

31   J. R. Aranzaes, M. C. Daniel and D. Astruc, *Can. J. Chem.*, 2006, **84**, 288–299.

32   L. Beer, R. W. Reed, J. L. Brusso, A. W. Cordes, R. C. Haddon, M. E. Itkis, K. Kirschbaum, D. S. MacGregor, R. T. Oakley and A. A. Pinkerton, *J. Am. Chem. Soc.*, 2002, **124**, 9498–9509.

33   L. Beer, J. F. Britten, J. L. Brusso, A. W. Cordes, R. C. Haddon, M. E. Itkis, D. S. MacGregor, R. T. Oakley, R. W. Reed and C. M. Robertson, *J. Am. Chem. Soc.*, 2003, **125**, 14394–14403.

34   K. Lekin, S. M. Winter, L. E. Downie, X. Bao, J. S. Tse, S. Desgreniers, R. A. Secco, P. A. Dube and R. T. Oakley, *J. Am. Chem. Soc.*, 2010, **132**, 16212–16224.

35   B. Odom, D. Hanneke, B. D'urso and G. Gabrielse, *Phys. Rev. Lett.*, 2006, **97**, 030801.

36   J. Zienkiewicz, P. Kaszynski and V. G. Young, *J. Org. Chem.*, 2004, **69**, 7525–7536.

37   H. M. McConnell, *J. Chem. Phys.*, 1956, **24**, 764–766.

38   D. R. Eaton, A. D. Josey, R. E. Benson, W. D. Phillips and T. L. Cairns, *J. Am. Chem. Soc.*, 1962, **84**, 4100–4106.

39   A. Hudson and K. D. J. Root, in *Advances in Magnetic and Optical Resonance*, Academic Press, 1971, vol. 5, pp. 1–79.

40   A. Bondi, *J. Phys. Chem.*, 1964, **68**, 441–451.

41   I. Dance, *New J. Chem.*, 2003, **27**, 22–27.

42   L. Beer, R. W. Reed, J. L. Brusso, A. W. Cordes, R. C. Haddon, M. E. Itkis, K. Kirschbaum, D. S. MacGregor, R. T. Oakley and A. A. Pinkerton, *J. Am. Chem. Soc.*, 2002, **124**, 9498–9509.

43   L. Beer, J. F. Britten, O. P. Clements, R. C. Haddon, M. E. Itkis, K. M. Matkovich, R. T. Oakley and R. W. Reed, *Chem. Mater.*, 2004, **16**, 1564–1572.

44   N. Zhang, S. Ayral-Kaloustian, T. Nguyen, R. Hernandez, J. Lucas, C. Discafani and C. Beyer, *Bioorganic Med. Chem.*, 2009, **17**, 111–118.

45   C. Bobbio, T. Rausis and M. Schlosser, *Chem. - A Eur. J.*, 2005, **11**, 1903–1910.

46   A. A. Leitch, X. Yu, C. M. Robertson, R. A. Secco, J. S. Tse and R. T. Oakley, *Inorg. Chem.*, 2009, **48**, 9874–9882.

47   G. R. Fulmer, A. J. M. Miller, N. H. Sherden, H. E. Gottlieb, A. Nudelman, B. M. Stoltz, J. E. Bercaw and K. I. Goldberg, *Organometallics*, 2010, **29**, 2176–2179.

48   C. P. Rosenau, B. J. Jelier, A. D. Gossert and A. Togni, *Angew. Chemie Int. Ed.*, 2018, **57**, 9528–9533.

49   C. F. Macrae, I. J. Bruno, J. A. Chisholm, P. R. Edgington, P. McCabe, E. Pidcock, L. Rodriguez-Monge, R. Taylor, J. van de Streek and P. A. Wood, *J. Appl. Crystallogr.*, 2008, **41**, 466–470.

50   Bruker AXS Inc., *APEX II*, 2007.

51   R. H. Blessing, *Acta Crystallogr. Sect. A*, 1995, **51**, 33–38.

52   C. B. Hübschle, G. M. Sheldrick and B. Dittrich, *J. Appl. Crystallogr.*, 2011, **44**, 1281–

1284.

53   G. M. Sheldrick, *Acta Crystallogr. Sect. A Found. Crystallogr.*, 2008, **64**, 112–122.

54   M. J. Frisch, G. W. Trucks, H. B. Schlegel, G. E. Scuseria, M. A. Robb, J. R. Cheeseman, G. Scalmani, V. Barone, B. Mennucci, G. A. Petersson, H. Nakatsuji, M. Caricato, X. Li, H. P. Hratchian, A. F. Izmaylov, J. Bloino, G. Zheng, J. L. Sonnenberg, M. Hada, M. Ehara, K. Toyota, R. Fukuda, J. Hasegawa, M. Ishida, T. Nakajima, Y. Honda, O. Kitao, H. Nakai, T. Vreven, J. Montgomery, J. A., J. E. Peralta, F. Ogliaro, M. Bearpark, J. J. Heyd, E. Brothers, K. N. Kudin, V. N. Staroverov, R. Kobayashi, J. Normand, K. Raghavachari, A. Rendell, J. C. Burant, S. S. Iyengar, J. Tomasi, M. Cossi, N. Rega, N. J. Millam, M. Klene, J. E. Knox, J. B. Cross, V. Bakken, C. Adamo, J. Jaramillo, R. Gomperts, R. E. Stratmann, O. Yazyev, A. J. Austin, R. Cammi, C. Pomelli, J. W. Ochterski, R. W. Martin, K. Morokuma, V. G. Zakrzewski, G. A. Voth, P. Salvador, J. J. Dannenberg, S. Dapprich, A. D. Daniels, O. Farkas, J. B. Foresman, J. V. Ortiz, J. Cioslowski and D. J. Fox, *Gaussian 09, Revis. D.01*, 2009.

Chapter 5

Three-Dimensional Magnetic Exchange Networks in Trigonal Bisdithiazolyl Radicals

5.1 Introduction

The discovery in the early 1990's of solid state ferromagnetic (FM) order in nitroxyl radicals[1-3] spawned research into the design of molecular magnetic materials based on the use of neutral, organic radicals as open-shell ($S = \frac{1}{2}$) building blocks.[4-8] However, in all light heteroatom radicals such as nitroxyls, verdazyls and triazinyls, intermolecular magnetic exchange interactions are weak, so that ordering temperatures T_C, when observed, are low (< 2 K).[9,10] With the inclusion of heavier heteroatoms such as sulfur into spin-bearing positions, stronger exchange interactions are possible,[11,12] although suppression of radical dimerization and consequent loss of magnetically active spins requires careful steric control.[13-16]

The pursuit of heavy atom radicals in which dimerization is inhibited more by electronic rather than steric factors, thereby facilitating more three-dimensional (3D) electronic and magnetic architectures, a critical requirement for bulk magnetic ordering, led to the development of resonance stabilized bisdithiazolyls (Figure 5.1a; **5-1**, R^2BPR^1).[17,18] In these radicals spin density is partitioned between two 1,2,3-dithiazole rings on either side of an N-alkylated pyridine bridge; the resulting singly occupied molecular orbital (SOMO) is shown in Figure 5.1b. When combined with the steric influence of the R^1/R^2 groups, the increased spin delocalization occasioned by the resonance effect is sufficient, in most cases,[19] to offset cofacial or pancake[20,21] dimerization, even when sulfur is replaced by its heavier congener selenium.[22] The resulting solid state structures consist of tilted rather than superimposed π-stacks of radicals locked into herringbone patterns (Figure 5.1c). Intermolecular magnetic interactions within these systems are dominated by pairwise isotropic exchange, often described in terms of the two-site single orbital Hubbard model.[23,24] Accordingly, the magnitude of the isotropic exchange energy J_{ij} between neighboring pairs (i,j) of radicals may be expressed as shown in Equation 5.1, where t_{ij} is the intermolecular hopping integral, directly related to the SOMO–SOMO overlap. U is the on-site Coulomb potential and K_{ij} is the electron exchange integral (the source of Hund's coupling). Qualitatively, strong overlap leads to a large virtual hopping term $4(t_{ij})^2/U$, which favours antiferromagnetic (AFM) exchange (-ve J), while FM exchange (+ve J) is favoured when SOMO–SOMO overlap is orthogonal, and hence t_{ij} nullified.

$$J_{ij} = 2K_{ij} - \frac{4t_{ij}^2}{U} \tag{5.1}$$

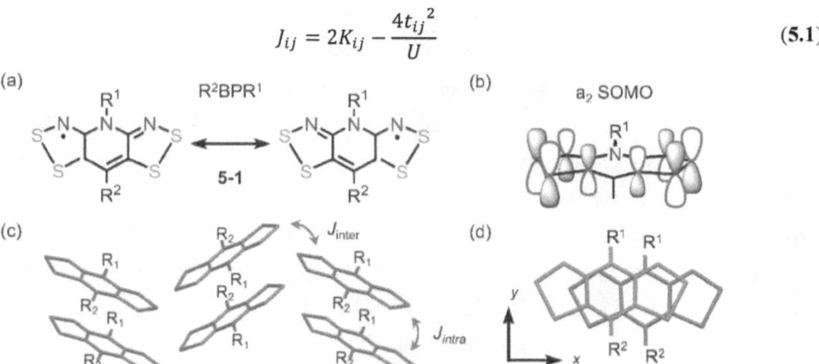

Figure 5.1 (a) Resonance structures and (b) a₂ symmetry SOMO of pyridine-bridged bisdithiazolyl radicals. (c) Herringbone packing of tilted π-stacks, with magnetic exchange interactions J_{intra} and J_{inter}. (d) Overlay of radicals along π-stacks, with local coordinate system to define slippage.

For descriptive purposes it is convenient to separate the magnetic interactions in these π-stacked structures into those present within the stacks (J_{intra}) and those developed between them (J_{inter}), as shown in Figure 5.1c. The magnitude of J_{intra} is strongly dependent upon the degree of tilting of the π-stacks, which affects SOMO–SOMO overlap and hence intersite hopping. Systematic studies using density functional theory broken symmetry (DFT-BS) methods to gauge the sign and magnitude of J_{intra} as a function of the slippage of radicals (Figure 5.1d) along the π-stacks have provided qualitative insight into the structural motifs that are conducive to FM versus AFM exchange.[25,26]

While such studies provide a useful conceptual framework for describing local exchange energies within π-stack arrays, bulk magnetic ordering, be it FM or AFM, requires magnetic interactions to span three dimensions.[27,28] In bisdithiazolyl radicals the design of structures possessing 2D and 3D magnetic exchange networks can be achieved through variation in the two beltline R^1/R^2 ligands. While these groups have little effect on molecular electronic properties (the a₂ SOMO is nodal at the site of substitution), their size plays a pivotal role in determining how the herringbone arrays interlock in the solid state. To illustrate this point, we show in Figure 5.2 the packing of several R^2BPR^1 radicals as a function of R^1 and R^2. In MeBPMe ($R^1 = R^2 = $ Me), space group $P2_1/c$ (Figure 5.2a), where the ligands are small and of the same size, neighboring herringbone arrays are aligned into corrugated ribbons separated by rows of interlocking methyl groups.[29] Regardless of the magnitude and sign of exchange coupling (J_{inter}) along the ribbons, the development of a 3D magnetic network is suppressed by the buffering effect of the ligands. When $R^1 = $ Et, $R^2 = $ H (Figure 5.2b) the space group remains unchanged, but the now dissimilar sizes of the ligands leads to weaving of the ribbons and the development of cross-links which introduce a third dimension for magnetic exchange.[30] In this case, the parent sulfur-based radical does not order, but its isostructural selenium-containing analogues do so as spin-canted antiferromagnets, with T_N (Néel ordering temperature) values reaching 28 K for full Se-incorporation.[31] In the R^1 = Et, R^2 = Cl derivative (Figure 5.2c) ribbon weaving is accentuated, and cross-links now occur around the $\bar{4}$ points of the tetragonal space group $P\bar{4}2_1/m$.[17] Here again the parent sulfur-based radical does not order, but with isostructural Se incorporation (partial or complete), bulk FM or spin-canted AFM materials are generated.[32] Many bulk ferromagnets belonging to this tetragonal family have since been prepared,[33–35] with T_C (Curie ordering temperature) reaching 27.5 K at 2.4 GPa for R^1 = Et, R^2 = I.[36] However, with lengthening of the R^1 alkyl chain (R^1 = *n*-Bu, Pn, Hx; R^2 = F) the tetragonal motif is lost and lateral interactions are confined to isolated pairs of π-stacks (Figure 5.2d); these materials are best described as spin ladders.[37] Thus, while the factors that control the overall preference for one phase over another are subtle, adjusting the size of the R^1/R^2 groups provides a useful "handle" for modifying supramolecular topology and hence magnetic dimensionality.

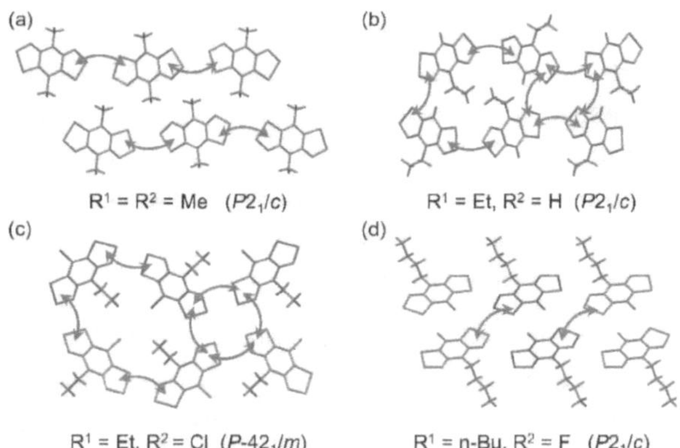

Figure 5.2 Dimensionality of magnetic interactions between π-stacks in bisdithiazolyl radicals R^2BPR^1. (a) In MeBPMe,[29] lateral magnetic interactions (red arrows) are confined to a single dimension. (b) In HBPEt [30], weaving of ribbons allows for magnetic cross-linking. (c) In ClBPEt,[17] cross-linking develops into pinwheel-like clusters about $\bar{4}$ points. (d) Longer R^1 groups, as in FBPBu,[37] lead to isolated pairwise-linked spin ladders.

Work on the two phenyl-substituted radicals PhBPR^1 (R^1 = Me, Et) was undertaken to examine the effect of large R^2 groups on structure and properties. Perhaps predictably, the combined buffering effect of the large R^1/R^2 groups in the R^2 = Et derivative (space group $P\bar{1}$) militated against a close approach of the herringbone ribbons, and 3D interactions were suppressed.[38] However, initial studies on the R^1 = Me (**5-2**) compound revealed a higher symmetry trigonal phase (space group $P3_121$), the structure of which allowed for cross-linking of neighbouring ribbons (Figure 5.3) and formation of helical arrays of intermolecular contacts about the 3_1 axes.[29] Magnetic measurements on this material and its isostructural Se-containing analogue provided no indication of FM interactions,[38] let alone FM magnetic order. The possibility of AFM order[39] or spin frustration,[40–43] features rarely observed in organic radicals and radical ions, was not explored. To address these latter issues, we have re-examined both the crystal structure and magnetic properties of PhBPMe. In so doing we have discovered a second polymorph, belonging to the orthorhombic space group $Pca2_1$ and hereafter referred to as the β-phase. The structure and magnetic properties of this β-phase provide an interesting contrast to those of the trigonal α-phase. Because of the high symmetry of the α-phase there are fewer unique exchange interactions than in the β-phase, and analysis of its magnetic susceptibility as a function of temperature using high temperature series expansion (HTSE) methods has provided insight into its magnetic structure. The experimental results for both phases have been interpreted in the light of exchange energies estimated by DFT-BS calculations.

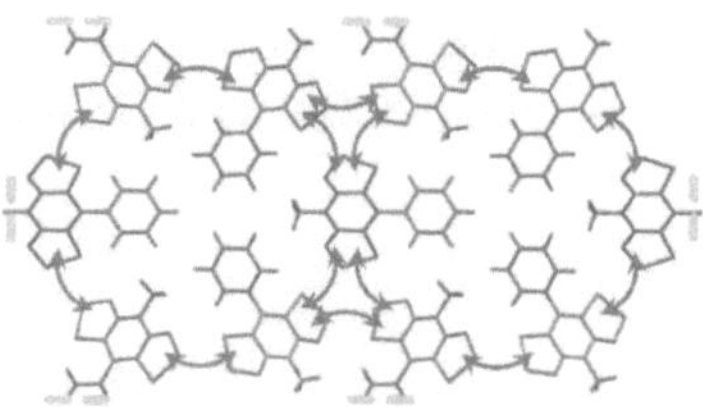

Figure 5.3 Packing of π-stacks in the α-phase (space group $P3_121$) of PhBPMe,[29] with helical arrays of intercolumnar exchange interactions (in red) about 3_1 axes.

5.2 Results and Discussion

5.2.1 Synthesis and Phase Separation of PhBPMe

Preparation towards the framework of PhBPMe (**5-2**) starts with alkylation of 2,6-dichloro-4-phenylpyridine (**5-3**) with methyl triflate, followed by amination to afford the N-methyl-2,6-diaminopyridinium salt (**5-4**; Scheme 5.1). Double Herz condensation[44] of the latter with sulfur monochloride in acetonitrile (MeCN) then yields **5-5**. Reduction of this material to the neutral radical can be achieved using octa- or decamethylferrocene (OcMFc or DcMFc). However, because of the polymorphic nature of the resulting radical, very precise procedures for reduction and crystal growth are required. Single crystals (fine needles) of the α-phase can be grown by codiffusion of solutions of **5-5** and OcMFc at ambient temperature, while bulk (microcrystalline) material suitable for magnetic measurements are obtained by the addition of solid OcMFc/DcMFc to a slurry of **5-5** in MeCN held at 0°C. If the temperature of this slurry is allowed to rise, even only to room temperature, the resulting microcrystalline product may be contaminated by the β-phase, the presence of which can be identified by infrared (IR) spectroscopy or, more definitively, by powder X-ray diffraction (PXRD). Bulk quantities of the β-phase suitable for both single-crystal diffraction and magnetic measurements can be grown by recrystallization of the as-reduced product (be it α-phase or mixed α-/β-phase) from hot, degassed 1,2-dichloroethane (DCE). Here too, however, small quantities of α-phase are sometimes apparent (by IR or PXRD analysis). As a result, samples of α- and β-phase materials used for magnetic measurements (*vide infra*) were always checked for phase purity (Figure 5.4).

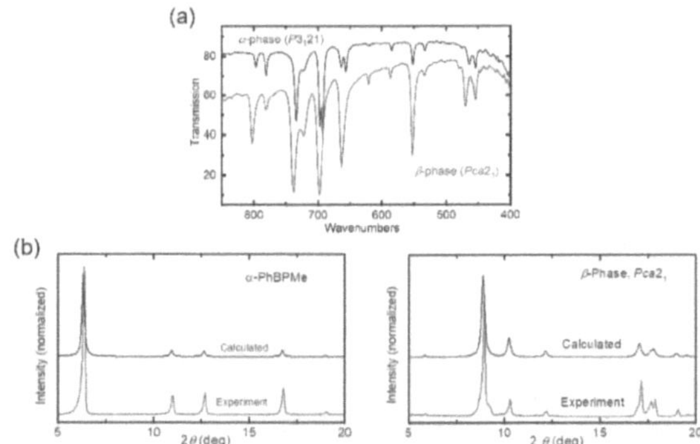

Scheme 5.1 Synthesis of PhBPMe. Reagents and conditions: (*i*) MeOTf, DCE, RT, 16 h; (*ii*) NH$_{3(g)}$, MeCN, 75°C, 24 h; (*iii*) S$_2$Cl$_2$ (excess), MeCN:DCE (1:10), reflux, 16 h; (*iv*) Oc-/DcMFc, MeCN, RT, 1 h.

Figure 5.4 (a) Calculated and observed PXRD patterns and (b) FTIR spectra of α- and β-phases for PhBPMe.

5.2.2 Structural Analysis of Polymorphic PhBPMe

The crystal structures of the two polymorphs of PhBPMe have been determined at ambient temperature and 100 K. Crystal data at both temperatures are summarized in Table A1.10 of Appendix A1, while ORTEP drawings of the asymmetric units, with atom numbering and pertinent intramolecular metrics (at 100 K), are provided in Figure 5.5. The near equality of these parameters in the two phases indicate similar molecular electronic structures. In both cases the molecular building blocks are slightly warped from planarity, a distortion defined here in terms of the dihedral angle ω between the planes of the two terminal SSN units. Warping is greater in the α-phase, where the molecules are twisted about a two-fold axis.

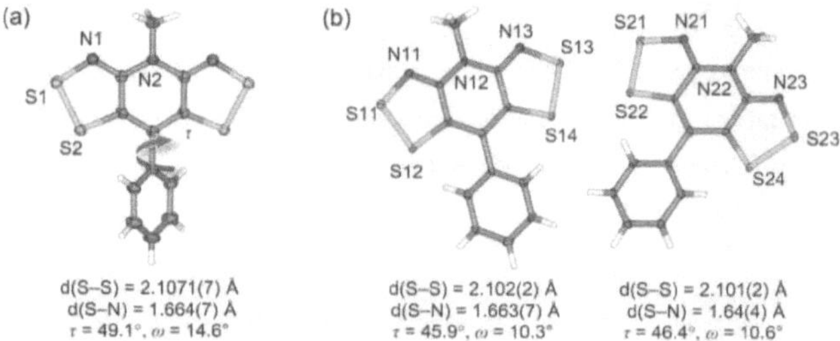

Figure 5.5 ORTEP drawings (50% thermal ellipsoids) of the asymmetric unit of (a) α- and (b) β-PhBPMe, with atom numbering and selected metrics (at 100 K) in Ångstroms. For averaged values, numbers in parentheses are the greater of the difference and the standard deviation. ω is the dihedral angle between the planes of opposite SSN units, and τ is the (mean) torsion angle for rotation of the phenyl rings. Disorder in the methyl protons of the α-phase is not shown.

The crystal structure of β-PhBPMe, orthorhombic space group $Pca2_1$, contains two independent radicals (1 and 2) in the asymmetric unit, each of which forms regularly spaced slipped π-stacks running parallel to the b-axis (Figure 5.6). Together these π-stacks pack in a herringbone pattern to afford corrugated ribbons which propagate parallel to the c-axis; adjacent stacks within these ribbons are connected by a series of close (inside the standard van der Waals separation)[45,46] intermolecular S···S′ contacts d_1–d_4. As a result of the combined steric influence of the beltline methyl/phenyl groups, which serve as buffers between neighboring ribbons, there are no close interstack S···S′ contacts parallel to the a-axis. This structural feature restricts the dimensionality of electronic and magnetic interactions of this phase.

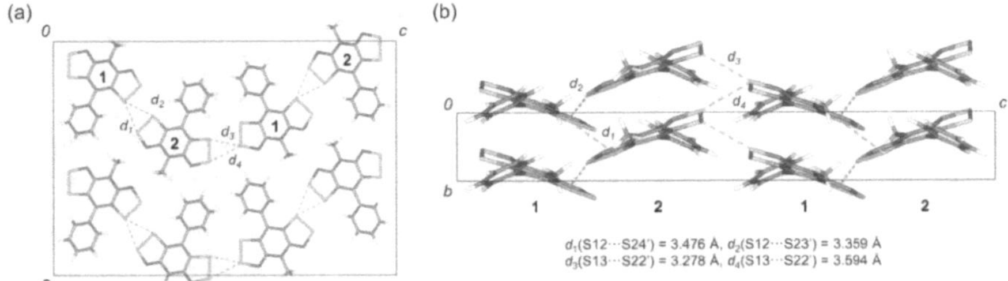

Figure 5.6 Unit cell projections of β-PhBPMe illustrating (a) the ribbons of π-stacked radicals propagating parallel to the c-axis, laced together by interstack contacts d_1–d_4 (shown in yellow dashed lines) and (b) a herringbone packing of slipped π-stacks. The two independent radicals are labeled 1 and 2. Colour code: C, grey; H, white; N, blue; S, yellow.

While corrugated herringbone arrays of slipped π-stacks are also found in α-PhBPMe, trigonal space group $P3_121$, the isolated-ribbon architecture found in the β-phase is replaced by a 3D network of lateral interstack interactions. The methyl and phenyl groups pack about 3_1 axes running parallel to the c-axis at $x = y = 0$ (Figure 5.7), while the radicals cluster into helical arrays along the 3_1 axes at $x = y = 1/3$. Because of its higher symmetry relative to the β-phase, there are only two unique intermolecular contacts d_1 and d_2, corresponding to interactions between radicals related by 120° (one-step) and 240° (two-step) rotations about the 3_1 axis. This simplification allows for a more in-depth analysis of the magnetic structure of this phase.

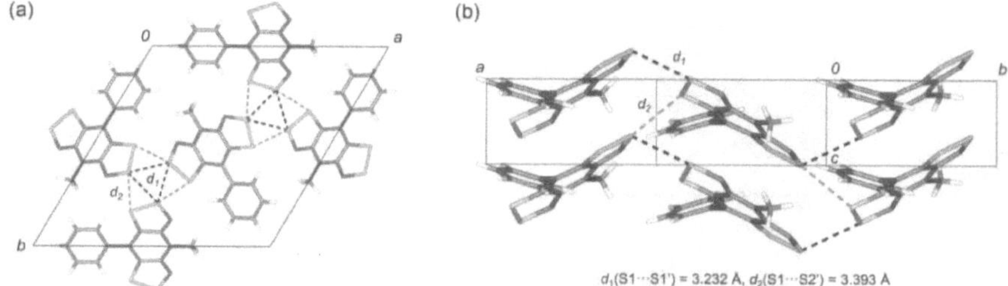

Figure 5.7 Unit cell projections of α-PhBPMe illustrating (a) π-stacked radicals clustering into helical arrays about 3_1 axes parallel to the c-axis, with interstack contacts d_1 and d_2 (shown in blue and green dashed lines, respectively) and (b) a herringbone packing of slipped π-stacks. Colour code: C, grey; H, white; N, blue; S, yellow.

5.2.3 Magnetic Measurements

Both phases of PhBPMe are Mott insulators,[47] with room temperature conductivity σ_{RT} values near 10^{-6} S cm⁻¹. Intermolecular interaction energies are therefore restricted to spin degrees of freedom, and accordingly the focus is placed on the magnetic properties of these materials. At the time, the polymorphism issue was not made aware, as a result of which the magnetic measurements were carried out on what was, in retrospect, a phase mixture.[29] Having established procedures for the separation of the α- and β-phases, the magnetic properties of both phases has been examined independently. The results of variable temperature DC magnetic susceptibility (χ) measurements collected over the range $T = 2–300$ K at field $H = 1000$ Oe, in the form of plots of χ and χT versus T, are presented in Figure 5.8. In both cases the effects of antiferromagnetic exchange interactions are apparent. For the α-phase a Curie-Weiss fit (Figure 5.8c) to the data from $T = 50–300$ K affords a C-constant of 0.389 emu K mol⁻¹, in good agreement with the value of 0.375 emu K mol⁻¹ expected for a radical with $S = \frac{1}{2}$ and a nominal $g = 2.0$; the associated value $\theta = -20.7$ K defines the scale for the magnitude of mean field AFM interactions. In the β-phase, AFM interactions are much stronger; a plot of χ versus T shows a low, broad maximum near $T = 150$ K with χT approaching 0.35 emu mol⁻¹ at $T = 300$ K.

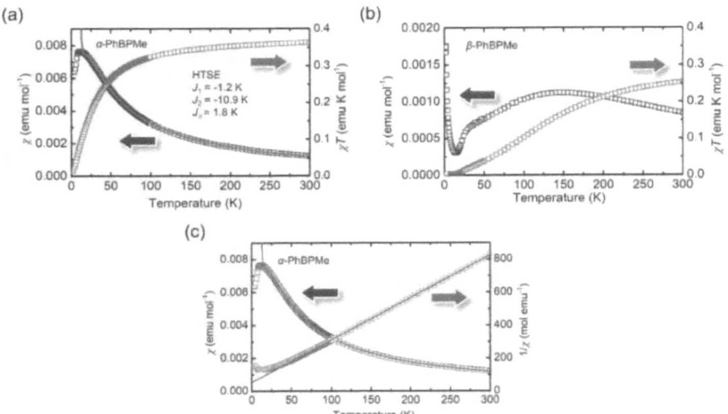

Figure 5.8 Cooling curve plots of the magnetic susceptibility χ (blue squares) and χT (green squares) versus T for (a) α- and (b) β-PhBPMe at a field $H = 1000$ Oe. For α-PhBPMe, the result of an HTSE fit to the data from 15 to 300 K using a J_1–J_2–J_π model is shown as a red line; as expected, the fit diverges below $T = 15$ K. (c) Cooling curve plots of the magnetic susceptibility χ (blue squares) and $1/\chi$ (green squares) versus T for α-PhBPMe at a field $H = 1000$ Oe.

To aid in the interpretation of the magnetic results in the two phases, we illustrate in Figure 5.9 the two unique nearest-neighbor interstack exchange pathways $J_{1,2}$ in the α-phase and the corresponding four independent interactions J_{1-4} in the β-phase. In addition, there is a single intrastack interaction J_π (J_{intra} above) in the α-phase and two such links in the β-phase (corresponding to coupling within the π-stacks of molecules 1 and 2). These exchange couplings, which we define here in magnitude and sign in terms of the phenomenological Hamiltonian in Equation 5.2, are defined geometrically with respect to the intermolecular contacts illustrated in Figure 5.6 and Figure 5.7. In the case of the β-phase, we have not attempted to extract from the magnetic data individual J-values based on a particular fit function. While the use of a Bonner-Fisher $S = \frac{1}{2}$ AFM chain model[48,49] based upon dominant interactions along the π-stacks is appealing, the presence of two independent chains (with two J_π values) militates against the effectiveness of this model.

$$\hat{\mathcal{H}}_{\text{ex}} = -2J_{ij}(\hat{\boldsymbol{S}}_i \cdot \hat{\boldsymbol{S}}_j) \tag{5.2}$$

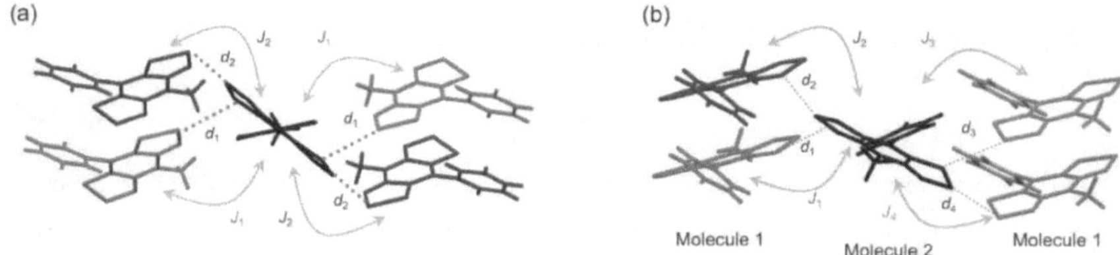

Figure 5.9 Interstack magnetic exchange interactions (a) $J_{1,2}$ in the α-phase and (b) J_{1-4} in the β-phase of PhBPMe.

While an AFM chain model could in principle be applied to model the susceptibility of the α-phase, we have elected to analyze its magnetic structure more deeply using high temperature series expansion (HTSE)[50] methods to extract values for each of the three interactions J_1, J_2 and J_π. This process required construction of the set of all pairwise interactions within a $6 \times 6 \times 6$ array of unit cells, following the scheme shown in Figure 5.10.

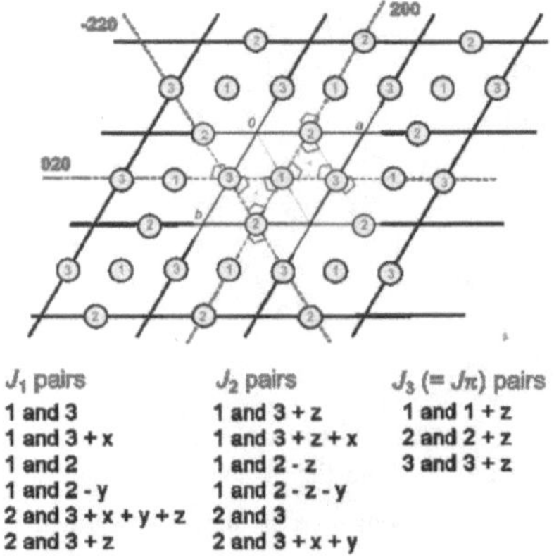

J_1 pairs	J_2 pairs	J_3 (= $J\pi$) pairs
1 and 3	1 and 3 + z	1 and 1 + z
1 and 3 + x	1 and 3 + z + x	2 and 2 + z
1 and 2	1 and 2 - z	3 and 3 + z
1 and 2 - y	1 and 2 - z - y	
2 and 3 + x + y + z	2 and 3	
2 and 3 + z	2 and 3 + x + y	

Figure 5.10 Pairwise coupling of radicals in $6 \times 6 \times 6$ cell of α-PhBPMe. The three radicals labelled (in red) 1, 2 and 3 constitute the trigonal unit cell, with radicals (in blue) in neighboring cells along x, y and z. Interacting pairs are designated by their translations, for example, $(1 + x)$ indicates radical 1 in the cell shifted once along the $+x$ direction.

The resulting function for χ, generated up to eighth order in $1/T$ (see Appendix A2), proved to be convergent over the range $T = 15$–300 K, the best fits being obtained for $J_1 = -1.2$ K, $J_2 = -10.9$ K, and $J_\pi =$

+1.8 K (Figure 5.8a). As expected from the initial Curie-Weiss analysis, AFM interactions dominate; application of the mean-field approximation (Equation 5.3)[51] with $zJ = 4J_1 + 4J_2 + 2J_\pi$ affords $\theta = -24$ K, in good agreement with the value (-20.7 K) obtained from the Curie-Weiss fit.

$$J = -\frac{3k_B\theta}{2zS(S+1)} \tag{5.3}$$

The experimentally derived J values, in particular the large AFM J_2, allow consideration of possible scenarios for bulk AFM ordering, a rarity in radical-based materials.[39] To this end we illustrate in Figure 5.11 the three spin spirals associated with the 3_1 axis of the $P3_121$ space group. One of these is based on J_1 interactions, which correspond to single steps about the spiral, while the other two are associated with two-step J_2 interactions. If J_1 were dominant, with both $J_2 \sim J_\pi \sim 0$, bulk AFM order could arise with a doubling of the unit cell c-dimension. However, the more likely situation, based on the HTSE fit, is that J_2 is dominant, that is $|J_2| >> |J_\pi| \sim |J_1|$, in which case the two independent spirals form the basis for a pair of extended AFM-aligned sublattices with cell dimension $4c$; one of these is illustrated in Figure 5.12. In this limit, the two interpenetrating, independent sublattices are each bipartite[52] and not frustrated, with the smallest loops consisting of six radicals. For this reason, one might expect the material to order magnetically with $T_N <<$ J_2.

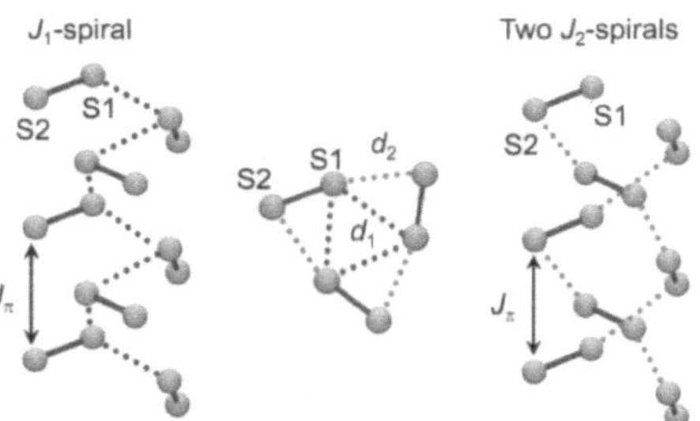

Figure 5.11 Three coaxial spin spirals about the 3_1 axes in α-PhBPMe based on J_1 and J_2 exchange interactions. For clarity, only two sulfur atoms in each radical are shown.

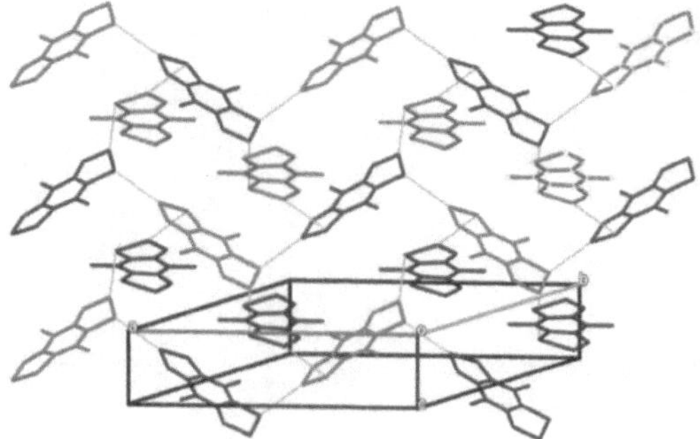

Figure 5.12 One of two 3D magnetic (bipartite) sublattices in α-PhBPMe based on dominant AFM J_2 exchange interactions. Color coding (red and blue) denotes sites of opposite spin.

The experimentally derived J values, in particular the large AFM J_2, allow consideration of possible scenarios for bulk AFM ordering, a rarity in radical-based materials.[39] To this end we illustrate in Figure 5.11 the three spin spirals associated with the 3_1 axis of the $P3_121$ space group. One of these is based on J_1 interactions, which correspond to single steps about the spiral, while the other two are associated with two-step J_2 interactions. If J_1 were dominant, with both $J_2 \sim J_\pi \sim 0$, bulk AFM order could arise with a doubling of the unit cell c-dimension. However, the more likely situation, based on the HTSE fit, is that J_2 is dominant, that is $|J_2| >> |J_\pi| \sim |J_1|$, in which case the two independent spirals form the basis for a pair of extended AFM-aligned sublattices with cell dimension $4c$; one of these is illustrated in Figure 5.12. In this limit, the two interpenetrating, independent sublattices are each bipartite[52] and not frustrated, with the smallest loops consisting of six radicals. For this reason, one might expect the material to order magnetically with $T_N <<$ J_2.

To explore the possibility of AFM ordering of the α-phase, we performed zero-field cooled/field-cooled (ZFC-FC) susceptibility measurements (Figure 5.13), but these showed no indication of a bifurcation in the ZFC and FC curves down to T = 2 K, even at H = 100 Oe. We note a very slight discontinuity at T = 8 K in both curves, which is accompanied by the development of a slight nonlinear variation in M versus H plots below 10 K, but subsequent heat capacity and AC susceptibility measurements (Figure 5.14) offered no support for a phase transition corresponding to 3D magnetic ordering. We suspect that these observations reflect the onset of short-range AFM correlations and that 3D ordering could occur at lower temperatures (T < 2 K)

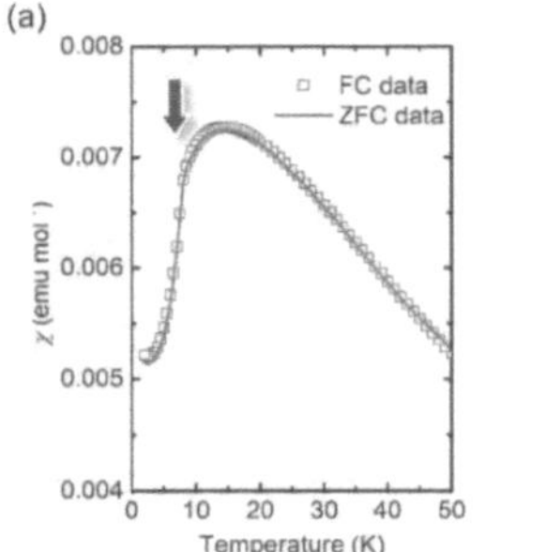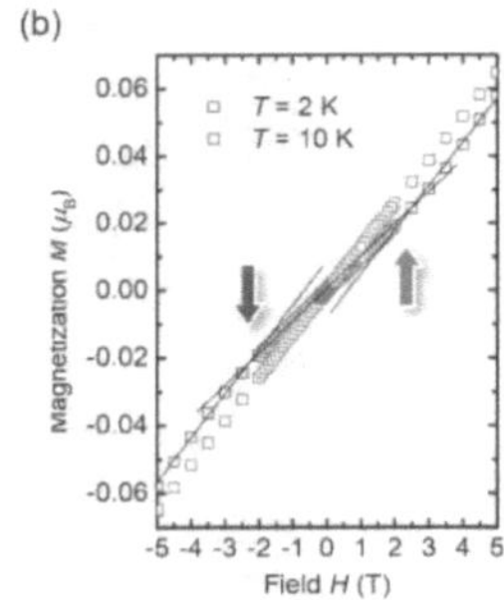

Figure 5.13 (a) ZFC (blue boxes) and FC (red line) magnetic susceptibility plots for α-PhBPMe as a function of temperature at a field $H = 100$ Oe. A slight discontinuity in both the ZFC and FC curves at 2 K is indicated with a green arrow. (b) Plots of M versus H for α-PhBPMe at 2 K (blue squares) and 10 K (red squares).

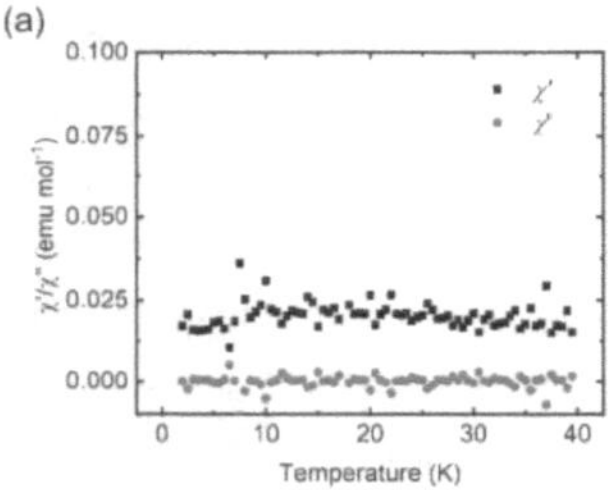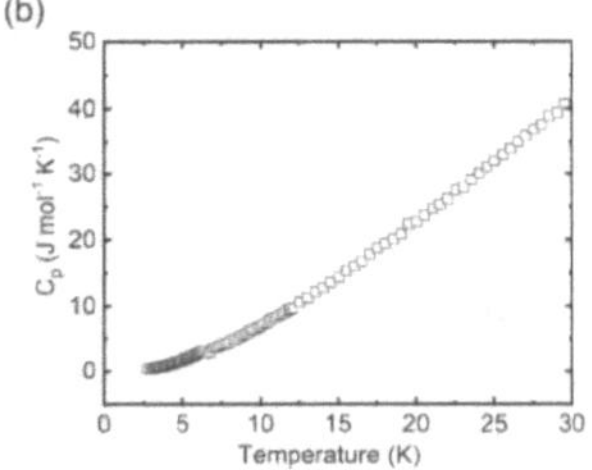

Figure 5.14 (a) AC magnetic susceptibility χ' and χ'' of α-PhBPMe at 10 Hz. (b) Heat capacity measurements on α-PhBPMe.

5.2.4 DFT Calculations

In order to place the experimental results within a theoretical context, DFT broken symmetry methods were used to estimate the pairwise magnetic exchange interactions shown in Figure 5.9. In this approach,[53–56] which has been successfully applied to a wide range of heterocyclic thiazyls[57,58] and selenazyls,[25,26] the exchange energy J_{ij} between two radicals is estimated from the total energies of the triplet (E_{TS}) and broken symmetry singlet (E_{BSS}) states and the respective expectation values $<S>$, according to Equation 5.4.

$$J = -\frac{(E_{TS} - E_{BSS})}{\langle S^2 \rangle_{TS} - \langle S^2 \rangle_{BSS}} \tag{5.4}$$

The results of a series of such calculations on both the α- and β-phases for PhBPMe, performed at the UB3LYP/6-311G(d,p) level using atomic coordinates taken from the 100 K crystal structures, are presented in Table 5.1. For the β-phase, three of the four interstack interactions J_{1-4} are predicted to be weakly FM, but both J_π-values are found to be strongly AFM, in accord with the large -ve θ value obtained from the magnetic measurements. In the case of α-phase the calculated J_π parameter is FM, but much more so than the value extracted from the HTSE analysis. For the two spiral interactions, J_2 is predicted to be AFM, in agreement with the experimental fit, while J_1 is found to be weakly FM, in contrast the weakly AFM value provided by the experiment. However, despite the individual discrepancies there is a reasonable overall correlation between theory and experiment, with the calculated values being uniformly more +ve (FM), suggesting that the DFT-BS approach is overestimating the effects of electron exchange (K_{ij}) effects relative to virtual hopping ($4(t_{ij})^2/U$), particularly so for J_π.

Table 5.1 UB3LYP/6-311G(d,p) exchange energies[a] for α and β-phases of PhBPMe.

	J_1 (K)	J_2 (K)	J_3 (K)	J_4 (K)	J_π (K)
α-phase	3.89	-6.21			34.87
β-phase	7.27	6.40	7.02	-0.22	-97.39 (1) -64.13 (2)

[a]J-values defined in Figure 5.9

The differences between the lateral exchange interactions $J_{1,2}$ in the α-phase and J_{1-4} in the β-phase are relatively small, but the J_π values in the two phases differ markedly. To explore this issue we carried out a series of model DFT-BS calculations at the UB3LYP/6-31G(d,p) level to track the J_π exchange interaction between a pair of idealized bisdithiazolyl radicals (with $R^1 = R^2 = H$) as a function of plate slippage, as defined in Figure 5.1d. In the trigonal space group $P3_121$ of α-PhBPMe, the radicals lie on two-fold axes, as a result of which the inclination of the radicals along the π-stacks is uniquely defined in terms of slippage in the local x-direction, with $y = 0$ by symmetry (Figure 5.15). Although there is no such symmetry-based restriction on the value of y in the β-phase, it is found to be near zero for both of the two independent molecules. A single parameter plot of J_π versus slippage along x thus provides qualitative insight into the magnitude of J_π in the two phases.[25,26]

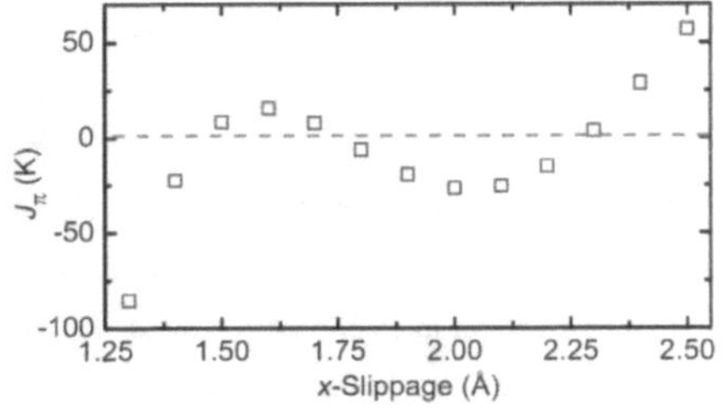

	α-Phase	β-Phase (1)	β-Phase (2)
δ, Å	3.493	3.635	3.605
x, Å	2.29	1.37	1.45
y, Å	0.00	0.09	0.07

Figure 5.15 UB3LYP/6-31G(d,p) calculated exchange interaction J_π for model 5-1 ($R^1 = R^2 = H$) dimers (plate separation $\delta = 3.5$ Å) as a function of lateral slippage x (with $y = 0$). Experimental slippage coordinates found in the α- and β-phases of PhBPMe are tabulated on the right.

The results reveal an oscillation in J_π as a function of x, a profile that reflects the subtle changes in SOMO–SOMO overlap and hence intermolecular hopping. Thus, when slippage is minimal (small x-values) the radicals are more nearly superimposed, resulting in strong AFM exchange.[25,26] This effect is seen to take hold, somewhat precipitously, below $x = 1.5$ Å, a finding that accounts for the strongly AFM J_π-values found for both molecules 1 and 2 in the β-phase, where $x = 1.37$ Å and 1.45 Å respectively. In the α-phase plate slippage is more pronounced ($x = 2.29$ Å), close to the region where the model calculations suggest a rapid crossover in J_π from AFM to FM values. Moreover, the apparent discrepancy between the single point calculations of J_π, which predict a much stronger FM interaction than that obtained from the HTSE fit, earlier ascribed to an over-emphasis of electron exchange relative to virtual hopping, can be manually rectified simply by shifting down the entire plot in Figure 5.15, in which case J_π values for both the α- and β-phases more closely match those obtained by experiment.

In principle, similar arguments could be applied to probe lateral exchange interactions, although geometrical models to track J-values as a function of structure are less obvious. For present purposes the J_π versus x plot provides predictive value in terms of the design of new trigonal radicals. For example, by fine-tuning of the size and shape of R^2 it may be possible to alter the degree of inclination of radicals along the π-stacks (the x-slippage value) and hence design structures which favour particular (AFM or FM) J_π values. This approach has certainly been successful in the design of magnetically active tetragonal radicals.[25,26,59]

5.3 Summary and Conclusion

Bisdithiazolyl radicals display a rich phase diagram as a function of the beltline ligands R^1 and R^2; triclinic, monoclinic, orthorhombic, tetragonal and trigonal cells have all been observed. In terms of their potential for magnetic ordering the higher symmetry space groups are appealing, as they afford packing patterns which promote the development of 3D intermolecular interactions. Phase control is, however, a major challenge, as demonstrated here for the R^1 = Me, R^2 = Ph derivative, which is polymorphic. That being said, the simplicity of the magnetic structure of the trigonal phase, coupled with the potential for modifying intermolecular exchange interactions through variations in the nature of the R^2 group, augurs well for the sculpting of new examples of this phase in which magnetic properties can be systematically tuned. The development of an HTSE function to extract the three unique exchange interactions J_1, J_2 and J_π from magnetic susceptibility data will facilitate the analysis of magnetic structure of these systems.

In the case of α-PhBPMe the dominance of the two-step spiral exchange interaction J_2 points to a 3D AFM magnetic network based on two equivalent bipartite sublattices which should order magnetically at $T < J_2$. However, while the onset of short-range AFM correlations near $T \sim 8$ K is apparent, there is no evidence from heat capacity, ZFC-FC and AC susceptibility measurements for 3D AFM ordering down to $T = 2$ K. Further studies on this material as well as new trigonal bisdithiazolyls and their selenium-based analogues should afford more detailed insight into the magnetic properties of this class of radicals.

5.4 Experimental Details

5.4.1 General Procedures

The starting material 8-phenyl-4-methyl-bis[1,2,3]dithiazolo[4,5-*b*:5',4'-*e*]pyridin-2-ium triflate **5-4** was prepared as previously described[29] (Scheme 5.1) by the condensation of *N*-methyl-2,6-diamino-4-phenyl-pyridinium triflate **5-4** with sulfur monochloride. The crude material was double recrystallized from acetic acid and MeCN prior to use. Octa- and decamethylferrocene (OcMFc and DcMFc) were obtained commercially and recrystallized from MeCN before use. The solvents MeCN and DCE were of reagent grade; MeCN was dried by distillation from P_2O_5 and CaH_2, and DCE by distillation from P_2O_5. Unless otherwise indicated all reactions were performed under an atmosphere of dry nitrogen. Infrared spectra were recorded either on a Nicolet Avatar FTIR spectrometer or a Varian 640 FTIR spectrometer with ATR. Phase purity of bulk samples of α- and β-PhBPMe was assessed by comparison of powder X-ray diffraction (PXRD) data with the patterns predicted from single-crystal X-ray diffraction experiments. All PXRD work was performed on a Rigaku Ultima IV diffractometer, with Cu K_α monochromatic radiation (λ = 1.54056 Å) and a θ–2θ geometry.

5.4.2 Experimental Procedures

Preparation of α-4-methyl-8-phenyl-bis[1,2,3]dithiazolo[4,5-*b*:5',4'-*e*]pyridin-2-yl (α-5-2). *Method 1: Single crystals for X-ray work.* Degassed solutions (four freeze-pump-thaw cycles) of OcMFc (0.085 g, 0.285 mmol) in MeCN (12 mL) and **5-5** (0.100 g, 0.212 mmol) in MeCN (12 mL) were allowed to diffuse together slowly at room temperature over a period of 16 h. The solvent was decanted to leave copper-colored needles of α-PhBPMe suitable for X-ray work. *Method 2: Bulk material for magnetic measurements.*

DcMFc (0.173 g, 0.530 mmol) was added to a slurry of **5-5** (0.250 g, 0.530 mmol) in degassed (4 freeze-pump-thaw cycles) MeCN (25 ml; and the mixture stirred for 1 h at 0°C. The resulting copper microcrystalline needles were collected by filtration, washed with 3×15 mL MeCN and dried in vacuo. Yield 0.143 g (0.444 mmol, 84%). IR: See Figure 5.4a for spectrum.

Preparation of β-4-methyl-8-phenyl-bis[1,2,3]dithiazolo[4,5-b:5′,4′-e]pyridin-2-yl (β-5-2). Decamethylferrocene (0.346 g, 1.06 mmol) was added to a solution of **5-5** (0.500 g, 1.06 mmol) in bubbled-degassed (11 mL) MeCN and stirred for 1 h at room temperature. The resulting dark purple solid was collected by filtration, washed $4 \times$ with 15 mL MeCN and dried in vacuo. Crude yield 0.312 g (0.968 mmol, 91%). Recrystallization from hot, degassed DCE (47 mL; 5 freeze-pump-thaw cycles) afforded metallic copper needle-like crystals. IR: See Figure 5.4a for spectrum.

5.4.3 Powder X-Ray Diffraction

PXRD measurements were performed on bulk powder samples of α- and β-PhBPMe using a Rigaku Ultima IV powder diffractometer with an X-ray source of Cu K_a ($\lambda = 1.5418$ Å) and a θ–2θ geometry. Phases purity of bulk samples was assessed by comparison of PXRD data with the patterns calculated from single-crystal data using the Mercury software program.[60]

5.4.4 Single-Crystal X-Ray Diffraction

Crystals of the α- and β-phases of PhBPMe were mounted on a MiTeGen MicroMount using Parabar oil. X-ray data on the α-phase were collected at 100 K on beamline I19 at the Diamond Light Source ($\lambda = 0.6889$ Å). Data for the β-phase were collected at 100 K *via* Cu radiation ($\lambda = 1.54178$ Å) on a Bruker Venture D8 dual-source micro-focus diffractometer fitted with a Photon 100 CMOS detector and at 293 K on beamline I19 at the Diamond Light Source ($\lambda = 0.6889$ Å). All data were collected using both φ and ω scans, and integration/reduction were undertaken with either Xia2[61–63] or Bruker ApexII[64] software. Subsequent computations were employed using the OLEX2 software interface.[65] Absorption corrections were applied to the data using SADABS.[64] Structures were solved *via* direct methods using SHELXT-2013[66] followed by refinement with SHELXL-2013.[67]

5.4.5 Magnetic Measurements

DC and AC magnetic susceptibility measurements were carried out over the range 2–300 K on a Quantum Design MPMS-XL SQUID magnetometer. Heat capacity measurements were performed on an Oxford Instruments MagLab EXA system using the relaxation method. Magnetic susceptibility data was corrected for diamagnetic contributions using Pascal's constants.[68,69] The High Temperature Series Expansion[50] (HTSE) function used in the analysis of the α-phase magnetic data is provided in Appendix A2.

5.4.6 Theoretical Calculations

DFT-BS exchange energy calculations on the individual exchange energies in α-PhBPMe (J_π, $J_{1,2}$) and in β-PhBPMe (J_π, J_{1-4}) were performed at the DFT level using the UB3LYP functional and the split-valence triple-basis set 6-311G(d,p), as contained in the Gaussian 09W[70] suite of programs. Tight convergence criteria were employed, and atomic coordinates were taken from the crystallographic data at 100 K. Exchange energies based on a Heisenberg Hamiltonian (Equation 5.2) for interacting pairs (i,j) of radicals were computed from Equation 5.3 using single-point energies of the triplet E_{TS} and broken symmetry singlet E_{BSS} states and their respective $\langle S^2 \rangle$ expectation values. Model 1D calculations (Figure 5.15) of J_π as a function of the π-stack slippage along the local x-axis were performed at the UB3LYP/6-31G(d,p) level using coordinates of a model R^2BPR^1 ($R^1 = R^2 = H$) obtained from a UB3LYP/6-31G(d,p) optimization in C_{2v} symmetry, as described elsewhere.[33–35]

5.5 References

1   M. Tamura, Y. Nakazawa, D. Shiomi, K. Nozawa, Y. Hosokoshi, M. Ishikawa, M. Takahashi and M. Kinoshita, *Chem. Phys. Lett.*, 1991, **186**, 401–404.

2   M. Kinoshita, *Philos. Trans. R. Soc. A*, 1999, **357**, 2855–2872.

3   R. Chiarelli, M. A. Novak, A. Rassat and J. L. Tholence, *Nature*, 1993, **363**, 147–149.

4   P. M. Lahti, in *Carbon Based Magnetism*, eds. T. Makarova and F. Palacio, Elsevier, 2006, pp. 23–52.

5   M. Kinoshita, in *π-Electron Magnetism: Magnetism from Molecules to Magnetic Materials*, eds. J. Veciana and D. Arcon, Springer Berlin Heidelberg, Berlin, Heidelberg, 2007, pp. 1–31.

6   I. Ratera and J. Veciana, *Chem. Soc. Rev.*, 2012, **41**, 303–349.

7   P. M. Lahti, *Adv. Phys. Org. Chem.*, 2011, **45**, 93–169.

8   H. Karoui, F. Le Moigne, O. Ouari and P. Tordo, in *Stable Radicals: Fundamentals and Applied Aspects of Odd-Electron Compounds*, ed. R. G. Hicks, John Wiley & Sons, Ltd, Chichester, UK, 2010, pp. 173–229.

9   T. Nogami, K. Tomioka, T. Ishida, H. Yoshikawa, M. Yasui, F. Iwasaki, H. Iwamura, N. Takeda and M. Ishikawa, *Chem. Lett.*, 1994, **23**, 29–32.

10   M. Mito, H. Nakano, T. Kawae, M. Hitaka, S. Takagi, H. Deguchi, K. Suzuki, K. Mukai and K. Takeda, *J. Phys. Soc. Japan*, 1997, **66**, 2147–2156.

11   A. J. Banister, N. Bricklebank, I. Lavender, J. M. Rawson, C. I. Gregory, B. K. Tanner, W. Clegg,

M. R. J. Elsegood and F. Palacio, *Angew. Chemie Int. Ed. English*, 1996, **35**, 2533–2535.

12   A. Alberola, R. J. Less, C. M. Pask, J. M. Rawson, F. Palacio, P. Oliete, C. Paulsen, A. Yamaguchi, R. D. Farley and D. M. Murphy, *Angew. Chemie Int. Ed.*, 2003, **42**, 4782–4785.

13   J. M. Rawson, A. Alberola and A. Whalley, *J. Mater. Chem.*, 2006, **16**, 2560.

14   J. M. Rawson, A. J. Banister and I. Lavender, in *Advances in Heterocyclic Chemistry*, ed. A. R. Katritzky, Elsevier, 1995, vol. 62, pp. 137–247.

15   D. A. Haynes, *CrystEngComm*, 2011, **13**, 4793–4805.

16   J. M. Rawson and C. P. Constantinides, in *Materials and Energy*, ed. J. S. Miller, World Scientific Publishing Co. Pte Ltd, 2018, vol. 4, pp. 95–124.

17   L. Beer, R. W. Reed, J. L. Brusso, A. W. Cordes, R. C. Haddon, M. E. Itkis, K. Kirschbaum, D. S. MacGregor, R. T. Oakley and A. A. Pinkerton, *J. Am. Chem. Soc.*, 2002, **124**, 9498–9509.

18   A. W. Cordes, R. C. Haddon and R. T. Oakley, *Phosphorus. Sulfur. Silicon Relat. Elem.*, 2004, **179**, 673–684.

19   A. A. Leitch, C. E. McKenzie, R. T. Oakley, R. W. Reed, J. F. Richardson and L. D. Sawyer, *Chem. Commun.*, 2006, 1088.

20   H. Z. Beneberu, Y. H. Tian and M. Kertesz, *Phys. Chem. Chem. Phys.*, 2012, **14**, 10713–10725.

21   K. E. Preuss, *Polyhedron*, 2014, **79**, 1–15.

22   J. L. Brusso, K. Cvrkalj, A. A. Leitch, R. T. Oakley, R. W. Reed and C. M. Robertson, *J. Am. Chem. Soc.*, 2006, **128**, 15080–15081.

23   C. J. Calzado, J. Cabrero, J. P. Malrieu and R. Caballol, *J. Chem. Phys.*, 2002, **116**, 2728–2747.

24   J. Huang and M. Kertesz, *J. Phys. Chem. A*, 2007, **111**, 6304–6315.

25   A. A. Leitch, X. Yu, S. M. Winter, R. A. Secco, P. A. Dube and R. T. Oakley, *J. Am. Chem. Soc.*, 2009, **131**, 7112–7125.

26   S. M. Winter, S. Hill and R. T. Oakley, *J. Am. Chem. Soc.*, 2015, **137**, 3720–3730.

27   Y. Shuku, Y. Hirai, N. A. Semenov, E. Kadilenko, N. P. Gritsan, A. V. Zibarev, O. A. Rakitin and K. Awaga, *Dalt. Trans.*, 2018, **47**, 9897–9902.

28   E. M. Fatila, A. C. Maahs, M. B. Mills, M. Rouzières, D. V. Soldatov, R. Clérac and K. E. Preuss, *Chem. Commun.*, 2016, **52**, 5414–5417.

29   L. Beer, J. F. Britten, O. P. Clements, R. C. Haddon, M. E. Itkis, K. M. Matkovich, R. T. Oakley and R. W. Reed, *Chem. Mater.*, 2004, **16**, 1564–1572.

30   L. Beer, J. F. Britten, J. L. Brusso, A. W. Cordes, R. C. Haddon, M. E. Itkis, D. S. MacGregor, R. T. Oakley, R. W. Reed and C. M. Robertson, *J. Am. Chem. Soc.*, 2003, **125**, 14394–14403.

31   A. A. Leitch, J. L. Brusso, K. Cvrkalj, R. W. Reed, C. M. Robertson, P. A. Dube and R. T. Oakley, *Chem. Commun.*, 2007, 3368.

32   C. M. Robertson, A. A. Leitch, K. Cvrkalj, R. W. Reed, D. J. T. Myles, P. A. Dube and R. T. Oakley, *J. Am. Chem. Soc.*, 2008, **130**, 8414–8425.

33   C. M. Robertson, D. J. T. Myles, A. A. Leitch, R. W. Reed, B. M. Dooley, N. L. Frank, P. A. Dube, L. K. Thompson and R. T. Oakley, *J. Am. Chem. Soc.*, 2007, **129**, 12688–12689.

34   C. M. Robertson, A. A. Leitch, K. Cvrkalj, D. J. T. Myles, R. W. Reed, P. A. Dube and R. T. Oakley, *J. Am. Chem. Soc.*, 2008, **130**, 14791–14801.

35   A. A. Leitch, K. Lekin, S. M. Winter, L. E. Downie, H. Tsuruda, J. S. Tse, M. Mito, S. Desgreniers, P. A. Dube, S. Zhang, Q. Liu, C. Jin, Y. Ohishi and R. T. Oakley, *J. Am. Chem. Soc.*, 2011, **133**, 6051–6060.

36   K. Lekin, K. Ogata, A. Maclean, A. Mailman, S. M. Winter, A. Assoud, M. Mito, J. S. Tse, S. Desgreniers, N. Hirao, P. A. Dube and R. T. Oakley, *Chem. Commun.*, 2016, **52**, 13877–13880.

37   K. Lekin, J. W. L. Wong, S. M. Winter, A. Mailman, P. A. Dube and R. T. Oakley, *Inorg. Chem.*, 2013, **52**, 2188–2198.

38   J. L. Brusso, S. Derakhshan, M. E. Itkis, H. Kleinke, R. C. Haddon, R. T. Oakley, R. W. Reed, J. F. Richardson, C. M. Robertson and L. K. Thompson, *Inorg. Chem.*, 2006, **45**, 10958–10966.

39   W. Fujita, K. Awaga, Y. Nakazawa, K. Saito and M. Sorai, *Chem. Phys. Lett.*, 2002, **352**, 348–352.

40   Y. Shimizu, T. Hiramatsu, M. Maesato, A. Otsuka, H. Yamochi, A. Ono, M. Itoh, M. Yoshida, M. Takigawa, Y. Yoshida and G. Saito, *Phys. Rev. Lett.*, 2016, **117**, 107203.

41   T. Itou, A. Oyamada, S. Maegawa, M. Tamura and R. Kato, *Phys. Rev. B*, 2008, **77**, 104413.

42   A. Mizuno, Y. Shuku, M. M. Matsushita, M. Tsuchiizu, Y. Hara, N. Wada, Y. Shimizu and K. Awaga, *Phys. Rev. Lett.*, 2017, **119**, 057201.

43   L. Postulka, S. M. Winter, A. G. Mihailov, A. Mailman, A. Assoud, C. M. Robertson, B. Wolf, M. Lang and R. T. Oakley, *J. Am. Chem. Soc.*, 2016, **138**, 10738–10741.

44   W. K. Warburton, *Chem. Rev.*, 1957, **57**, 1011–1020.

45   A. Bondi, *J. Phys. Chem.*, 1964, **68**, 441–451.

46   I. Dance, *New J. Chem.*, 2003, **27**, 22–27.

47   N. F. Mott, *Proc. Phys. Soc. Sect. A*, 1949, **62**, 416.

48   J. C. Bonner and M. E. Fisher, *Phys. Rev.*, 1964, **135**, A640–A658.

49   D. C. Johnston, R. K. Kremer, M. Troyer, X. Wang, A. Klümper, S. L. Bud'ko, A. F. Panchula and P. C. Canfield, *Phys. Rev. B*, 2000, **61**, 9558–9606.

50   H.-J. Schmidt, A. Lohmann and J. Richter, *Phys. Rev. B*, 2011, **84**, 104443.

51   C. Kittel, *Introduction to Solid State Physics*, John Wiley & Sons, New York, 5th edn., 2004.

52   E. H. Lieb, *Phys. Rev. Lett.*, 1989, **62**, 1201–1204.

53   L. Noodleman, *J. Chem. Phys.*, 1981, **74**, 5737–5743.

54   L. Noodleman and E. R. Davidson, *Chem. Phys.*, 1986, **109**, 131–143.

55   H. Nagao, M. Nishino, Y. Shigeta, T. Soda, Y. Kitagawa, T. Onishi, Y. Yoshioka and K. Yamaguchi, *Coord. Chem. Rev.*, 2000, **198**, 265–295.

56   M. Deumal, M. A. Robb and J. J. Novoa, *Prog. Theor. Chem. Phys.*, 2007, **16**, 271–289.

57   R. I. Thomson, C. M. Pask, G. O. Lloyd, M. Mito and J. M. Rawson, *Chem. - A Eur. J.*, 2012, **18**, 8629–8633.

58   M. Deumal, J. M. Rawson, A. E. Goeta, J. A. K. Howard, R. C. B. Copley, M. A. Robb and J. J. Novoa, *Chem. - A Eur. J.*, 2010, **16**, 2741–2750.

59   K. Lekin, K. Ogata, A. Maclean, A. Mailman, S. M. Winter, A. Assoud, M. Mito, J. S. Tse, S. Desgreniers, N. Hirao, P. A. Dube and R. T. Oakley, *Chem. Commun.*, 2016, **52**, 13877–13880.

60   C. F. Macrae, I. J. Bruno, J. A. Chisholm, P. R. Edgington, P. McCabe, E. Pidcock, L. Rodriguez-Monge, R. Taylor, J. Van De Streek and P. A. Wood, *J. Appl. Crystallogr.*, 2008, 41, 466–470.

61   Collaborative Computational Project Number 4, *Acta Crystallogr. Sect. D Biol. Crystallogr.*, 1994, **50**, 760–763.

62   P. Evans, *Acta Crystallogr. Sect. D Biol. Crystallogr.*, 2006, **62**, 72–82.

63   G. Winter, *J. Appl. Crystallogr.*, 2010, **43**, 186–190.

64   Bruker AXS Inc., *APEX II*, 2007.

65   O. V. Dolomanov, L. J. Bourhis, R. J. Gildea, J. A. K. Howard and H. Puschmann, *J. Appl. Crystallogr.*, 2009, **42**, 339–341.

66   G. M. Sheldrick, *Acta Crystallogr. Sect. A Found. Adv.*, 2015, **71**, 3–8.

67   G. M. Sheldrick, *Acta Crystallogr. Sect. C Struct. Chem.*, 2015, **71**, 3–8.

68   R. L. Carlin, *Magnetochemistry*, Springer Berlin Heidelberg, Berlin, Heidelberg, 1986.

69   G. A. Bain and J. F. Berry, *J. Chem. Educ.*, 2008, **85**, 532–536.

70   M. J. Frisch, G. W. Trucks, H. B. Schlegel, G. E. Scuseria, M. A. Robb, J. R. Cheeseman, G. Scalmani, V. Barone, B. Mennucci, G. A. Petersson, H. Nakatsuji, M. Caricato, X. Li, H. P. Hratchian, A. F. Izmaylov, J. Bloino, G. Zheng, J. L. Sonnenberg, M. Hada, M. Ehara, K. Toyota, R. Fukuda, J. Hasegawa, M. Ishida, T. Nakajima, Y. Honda, O. Kitao, H. Nakai, T. Vreven, J. Montgomery, J. A., J. E. Peralta, F. Ogliaro, M. Bearpark, J. J. Heyd, E. Brothers, K. N. Kudin, V. N. Staroverov, R. Kobayashi, J. Normand, K. Raghavachari, A. Rendell, J. C. Burant, S. S. Iyengar, J. Tomasi, M. Cossi, N. Rega, N. J. Millam, M. Klene, J. E. Knox, J. B. Cross, V. Bakken, C. Adamo, J. Jaramillo, R. Gomperts, R. E. Stratmann, O. Yazyev, A. J. Austin, R. Cammi, C. Pomelli, J. W. Ochterski, R. W. Martin, K. Morokuma, V. G. Zakrzewski, G. A. Voth, P. Salvador, J. J. Dannenberg, S. Dapprich, A. D. Daniels, O. Farkas, J. B. Foresman, J. V. Ortiz, J. Cioslowski and D. J. Fox, *Gaussian 09, Revis. D.01*, 2009.

Chapter 6

Conclusion and Future Directions

6.1 Conclusive Statements

Since the discovery of the $(SN)_x$ polymer, heterocyclic thiazyl radicals have taken a spotlight in the development of new materials and devices owing to their promising optical, magnetic and-electronic properties. The array of five- and six-membered ring systems are appealing classes of stable-radical building blocks, from which judicious choice of peripheral substituents can provide a platform to tune the physical properties of the radical at the molecular level. Moreover, the substituents can open a gateway to direct the crystal packing, and ultimately, create multi-dimensional networks for electronic communication. With these synthetic handles, efforts to crystal engineer the assembling of radicals in the solid state by molecular design remain a fundamental interest amongst researchers in this field.

The number of possible variances to thiazyl radicals are essentially limitless; however, the construction of targeted frameworks is often presented with synthetic challenges. This was realized in our exploration of thienyl- and pyridyl-substituted 1,2,4,6-thiatriazinyls (TTAs); the first examples of TTAs functionalized *via* heteroaromatic substituents. The preparation of each derivative is heavily dependent on the nature of the aromatic moiety, from the generation of the imidoylamidine precursors to the purification of neutral radical. Many of the synthetic hurdles stem from the increased basicity and nucleophilicity of pyridyl nitrogen atoms compared to the sulfur atoms of thiophene, as illustrated by the proclivity of the pyridyl substituents to react non-innocently with sulfur halides and acids. Nonetheless, once we established the synthesetic methodologies and isolated the radicals, characterization of the two functionalized TTAs clearly demonstrated the ability of the heteroaryl rings to influence both the molecular and solid-state properties. Spectroscopic studies revealed the contrasting effects on the electronic structure (e.g., electron-releasing vs. withdrawing) by these substituents, while crystal analysis emphasized the capacity to which the additional heteroatoms play a role in the lattice assembly of molecules. In addition, the exocyclic groups illustrated an enhancement of intermolecular contacts between cofacially paired radicals, both in strength and numbers when juxtaposed to their phenyl analogue, from which the radical systems were not exempted from π-dimerization.

Despite the lost of spin properties in the solid state, the framework of these radicals present opportunities for metal coordination in solution that may stabilize the radical character. In addition, the fixed tridentate pocket provides opportunities for strengthening ligand-metal interactions that become beneficial for magnetic communication. Given the redox-flexibility of the TTA ligand, we investigated lanthanide coordination chemistry utilizing the pyridyl-substituted 8π-TTA precursor, a more stable, robust form of the ligand framework. Complexation resulted in crystals comprised of a dinuclear Dy^{III} system, of which the sulfur atom of the TTA framework is oxidized and the molecules are converted into bridging ligands. To provide insights on the self-assembly process, we performed ^1H NMR studies on an isostructural yttrium analogue, from which the formation of intermediates, both closed- and open-shell, could be probed through the pyridyl hydrogen atoms, given the diamagnetic metal center and symmetrical ligand framework. The basis of the results infers an *in situ* metal-assisted ligand oxidation followed by dimerization to form discrete dinuclear molecules. Moreover, magnetic measurements on the paramagnetic Dy^{III} complex conveyed the presence of single-molecule magnet behaviour, attributed to the single-ion anisotropy of the dysprosium ion.

The benefits of incorporating thienyl moieties to the TTA framework expand further than modulating the properties of thiazyls. These functional groups do not interfere upon ring closure with the insertion of sulfur, which, from a synthetic standpoint, provides an efficient route to the neutral radical. In light of this, we extended our work on thienyl-functionalization to the pyridine-bridged bisthiadiazinyl (bis-TDA) radicals on the basis of unveiling heteroaromatic effects on a resonance-stabilized open-shell framework. A set of radicals were synthesized that additionally probed the structure-property relationship from the inclusion of halogens to the molecular skeleton, given the multiple sites for functionalization. EPR studies revealed a rich hyperfine splitting pattern that reflected the distribution of the spin density along the extended π-substituents, while UV-Vis and electrochemical studies highlighted the ability for these substituents to tailor the energies of the frontier molecular orbitals. Structural analysis of the polymorphs unveiled the influence of the thienyl rings towards directing the solid-state packing through its flexibility to engage in favourable π–π and electrostatic interactions. On the other hand, implementing halogens to the thiazyl framework provided a tool to modify the alignment of molecules supramolecularly.

Polymorphism, in and of itself, is not uncommon in thiazyl radicals. In some instances, the structural differences between the phases can result in significant changes to the bulk properties of the material, as was the case when we sought to separate the low-temperature (LT) and high-temperature (HT) phases of a bisdithiazolyl radical. The two polymorphic structures differ substantially; the HT phase, belonging to an orthorhombic space group, consists of ribbon-like arrays of stacked molecules, while the LT phase belongs to a rare, higher-symmetry, trigonal space group and is composed of a honeycomb structure of stacked radicals. Similar to the bis-TDAs, meticulous attention to the conditions of both the reaction and purification stages were necessary to isolate each phase selectively. In so doing, magnetic measurements on the high-symmetry LT phase reflected the transition of an AFM ordered state below 8 K. To interpret the data, we developed of a high-temperature series expansion function to model the magnetic behaviour, from which we extracted three unique exchange interactions arising in the system. On the basis of these values, the AFM phase can be seen comprised of two equivalent spin-spiral sublattices spanning in three dimensions.

6.2 Future Directions

As this book comes to an end, there have many avenues that continue to flourish from this research that are currently ongoing in the group. Synthetically, we have reported the groundwork towards the functionalization of TTA radicals with N-heterocycles as described in Chapter 2, from which a range of derivatives can be envisioned for the design of novel redox-active ligand systems in coordination chemistry. In particular, we have pursued the synthesis of the 2-pyrimidyl analogue (**6-1**; Scheme 6.1) for the development of extended coordination polymers and networks, given the multiple binding pockets to bridge metal centers. The spin-bearing framework enables opportunities to investigate exchange pathways between various metal ions, which may lead to overall enhancement of the magneto-anisotropy in multinuclear complexes. The synthetic route towards **6-1** follows suit with that of the pyridyl derivative starting with the imidoylamidine **6-5** we previously reported,[1] from which we have isolated and characterized the 8π-TTA precursor **6-12** *via* SCXRD (Figure 6.1a). Attempts to oxidize **6-12** have been challenging, from what we believe stems from the reduced solubility of the material and high proton affinity along the TTA ring.

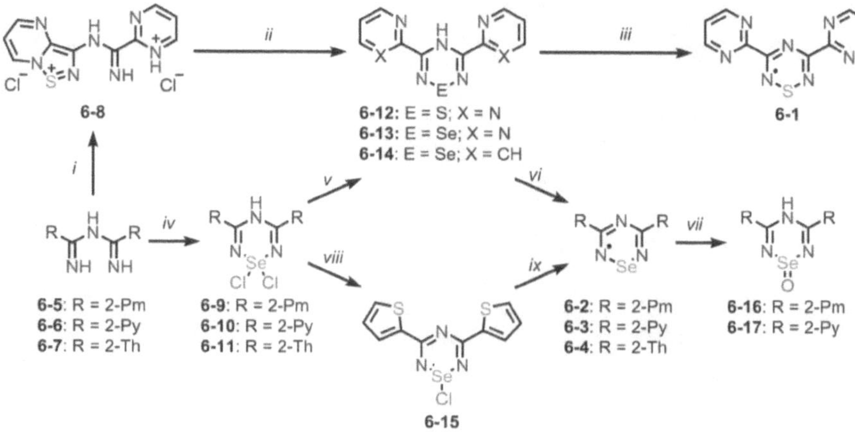

Scheme 6.1 Synthesis of selected TTAs and STAs. Reagents: (*i*) S_2Cl_2; (*ii*) PhCl; (*iii*) oxidant, base; (*iv*) SeCl$_4$; (*v,*) SbPh$_3$, base; (*vi*) NCS, base; (*vii*) O$_2$; (*viii*) base; (*iv*) PhSb$_3$.

Alternatively, selenium can replace sulfur in the TTA framework to afford 1,2,4,6-selenatriazinyls (STAs).[2] The benefits of selenium incorporation, in regards to spin-bearing ligands, are expected to come from the spin-orbit coupling and the enhanced interactions between magnetic orbitals. With the imidoylamidine precursors (**6-5–6-7**) in hand, we have worked towards heteroaromatic-functionalized STAs **6-2–6-4** as described in Scheme 6.1, with single-crystal X-ray analysis supporting structural identification of intermediates provided in Figure 6.1b–f. For each derivative, the formation of the STA framework comes from the treatment of the imidoylamidine with SeCl$_4$ to generate **6-9–6-11**. In the case of the *N*-heteroaryl analogues, reduction of the central framework in presence of base affords the removal of chloro substituents as in **6-13** and **6-14**. Subsequent oxidation of these intermediates and isolation of both neutral radicals (**6-2** and **6-3**) have come with challenges. While we believe the radical is being produced upon oxidation, we have only acquired crystals of the decomposed oxides (**6-16** and **6-17**), which in turn suggests the radical product is being formed. When applying these reaction conditions to the thienyl derivative, treatment of **6-11** with triphenylantimony in the presence of base results in the neutral radical, which dimerizes in the solid state (Figure 6.1g). We suspect that the mechanism involves the intermediate **6-15**, in which **6-11** is deprotonated and subsequently aromatized through the displacement of a chlorine atom, and from then can be reduced by triphenylantimony. From our work, we anticipate completing the series of STA radicals in the near future and exploring metal coordination with these systems.

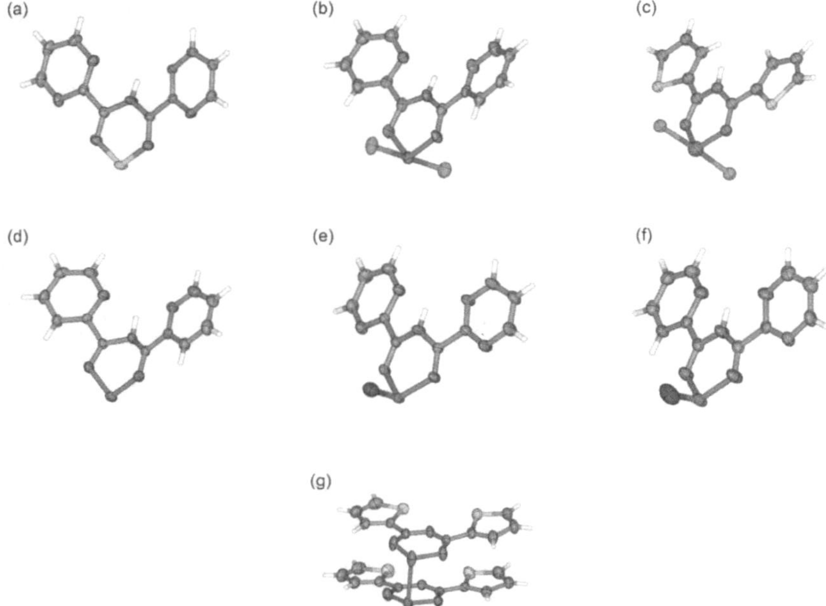

Figure 6.1 ORTEP drawings (50% thermal ellipsoids) of the asymmetric unit of (a) **6-12** (b) **6-10** (c) **6-11** (d) **6-14** (e) **6-16** (f) **6-17** and (g) at 200 K. Colour code: C, grey; H, white; N, blue; S, yellow; Cl, green; Se, orange.

Another current area of research is expanding from the work in Chapter 4. Intrigued by the polymorphic behaviour of **6-18** (R = H, F, Cl; Chart 6.1), we are currently investigating the magnetic properties of these systems and exploring the possibility for temperature-dependent solid-solid phase transitions that may bring to light bistable regimes for molecular switching applications. Collaborative work has been initiated on measuring the magnetic behaviour of the isolated phases of **6-18** (R = H, F, Cl), while DFT studies are being performed to provide insights on the degree of exchange interactions between these frameworks. Additionally, we are also exploring N-heteroaromatic functionalization and selenium replacement of these frameworks (**2-19–2-21** in Chart 6.1), which has never been reported for this class of thiazyl radicals. On the basis of our work in Chapter 2 and 4, we are expecting the addition of the pyridyl and pyrimidyl attachments to direct the crystal packing of molecules into closer arrangements, which may lead to improvements on the magnetic and conductive properties in thiazyl materials. The precursors for the N-heterocyclic derivatives have been isolated and characterized. As expected though, the engagement of these groups upon treatment with sulfur halides complicates the characterization of the products afforded by in the reaction. Efforts to circumvent these issues are underway.

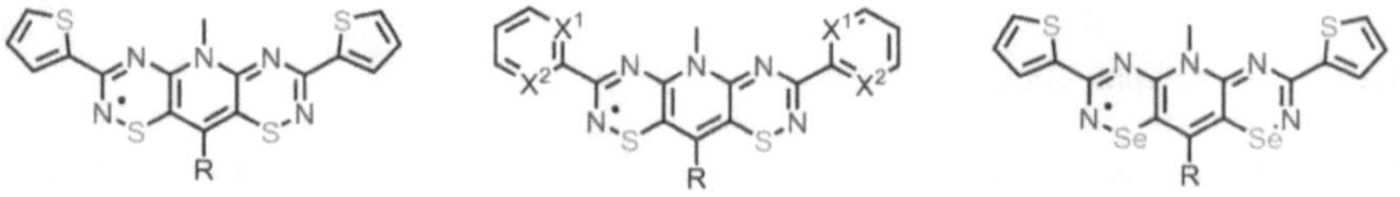

6-18

6-19: $X^1 = N$; $X^2 = CH$
6-20: $X^1 = N$; $X^2 = N$

6-21

Chart 6.1 Molecular variants of the bis-TDA radical for the study of structure-property relationships.

Lastly, the magnetic properties from the trigonal bis-DTA radical system (**6-22**; Chart 6.2) have the potential to be enhanced by full selenium incorporation into the molecular framework. We expect crystals of **6-23** (Chart 6.2) to be isostructural to its sulfur counterpart, from which the exchange interactions between radical centers can be strengthened due to increase SOMO–SOMO overlap. Furthermore, the simplicity of the magnetic architecture presents opportunities to systematically tune the magnetic properties by structural modifications. For example, this may come through the employment of aryl groups with a similar size to the phenyl ring (**6-24** in Chart 6.2) or possibly switching the position of the groups across the central axis of the molecule (**6-25** in Chart 6.2). In that regard, synthetic preparations towards these modifications are at this time being researched.

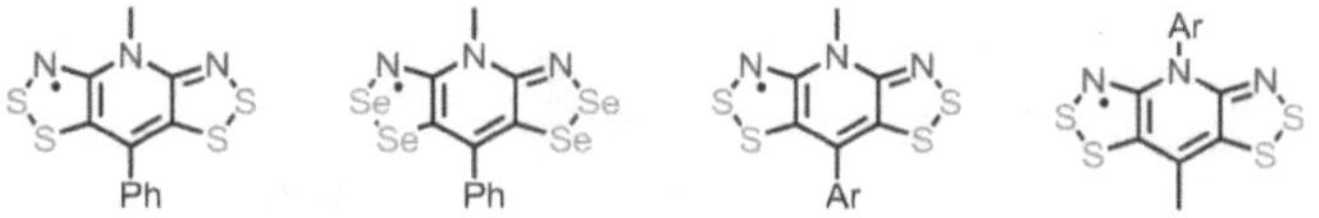

6-22   **6-23**   **6-24**   **6-25**

Chart 6.2 Molecular variants of the bis-DTA radical for the study of structure-property relationships.

6.3 References

1   M. Yousaf, N. J. Yutronkie, R. Castañeda, J. A. Klein and J. Brusso, *New J. Chem.*, 2017, **41**, 12218–12224.

2   R. T. Oakley, R. W. Reed, A. W. Cordes, S. L. Craig and J. B. Graham, *J. Am. Chem. Soc.*, 1987, **109**, 7745–7749.

Appendix

A.1 Crystallographic Data

Table A1.1 Crystallographic data and selected data collection parameters for **2-12–2-15** in Chapter 2.

	2-12	2-13	2-14	2-15
Formula	$C_{12}H_{10}Br_5N_5S_1$	$C_{31}H_{27}Br_5N_{12}S_2$	$C_{12}H_9N_5OS_1$	$C_{10}H_7N3OS_3$
Formula Weight	655.86	1031.31	271.30	281.37
Crystal System	Monoclinic	Triclinic	Triclinic	Monoclinic
Space Group	$P2_1/c$	$P\bar{1}$	$P\bar{1}$	$P2_1$
a (Å)	13.5657(3)	9.9574(3)	6.4045(3)	9.3997(12)
b (Å)	7.7176(2)	12.1849(3)	7.6170(4)	4.2844(5)
c (Å)	17.8819(4)	16.3791(4)	12.4324(6)	14.2053(16)
α (°)	90	110.4818(16)	82.682(2)	90
β (°)	100.7904(15)	92.7861(17)	82.181(2)	91.627(4)
γ (°)	90	97.4104(17)	76.697(2)	90
V (Å^3)	1839.04(8)	1836.67(9)	581.82(5)	571.85(12)
Z	2	2	2	2
ρ_{calc} (g·cm^3)	2.369	1.865	1.549	1.634
T (K)	200(2)	200(2)	200(2)	200(2)
μ (mm^{-1})	11.044	5.627	0.277	0.632
$2\Theta_{max}$ (°)	28.302	28.322	28.246	28.276
Total Reflections	25235	26792	4276	5493
Unique Reflections	4501	9017	2831	2617
R$_1$, wR$_2$ (on F^2)	0.0388, 0.0957	0.0347, 0.0721	0.0575, 0.1618	0.0419, 0.0841

Table A1.2 Selected bond lengths[a] from the asymmetric units of **2-12**–**2-15** in Chapter 2.

	2-12	2-13	2-14	2-15
S1–N1 (Å)	1.603(4)	1.594(2) 1.589(2)[b]	1.681(3)	1.645(3)
N1–C1 (Å)	1.340(5)	1.326(3) 1.348(3)[b]	1.285(4)	1.306(5)
C1–N2 (Å)	1.338(5)	1.346(4) 1.328(4)[b]	1.373(4)	1.366(5)
N2–C2 (Å)	1.345(6)	1.339(3) 1.343(3)[b]	1.367(4)	1.322(4)
C2–N3 (Å)	1.318(5)	1.338(3) 1.336(3)[b]	1.283(4)	1.341(5)
N3–S1 (Å)	1.622(4)	1.595(2) 1.606(2)[b]	1.687(3)	1.704(3)
S1–Br1 (Å)	2.347(1)	2.453(1) 2.411(1)[b]		
S1–O1 (Å)			1.459(4)	1.468(3)

[a] Atom numbering based of Figure 2.1 and differs from crystallographic labelling. [b] Bond length from second molecule in asymmetric unit.

Table A1.3 Crystallographic data and selected data collection parameters for **2-2** and **2-3** in Chapter 2.

	2-2 (Py$_2$TTA)	**2-3** (Th$_2$TTA)
Formula	C$_{24}$H$_{16}$N$_{10}$S$_2$	C$_{20}$H$_{12}$N$_6$S$_6$
Formula Weight	508.59	528.72
Crystal System	Triclinic	Monoclinic
Space Group	$P\bar{1}$	$P2_1/n$
a (Å)	8.3797(3)	11.3840(6)
b (Å)	10.5524(3)	10.7358(6)
c (Å)	13.0609(4)	17.9318(9)
α (°)	81.4929(12)	90
β (°)	84.0073(13)	98.461(3)
γ (°)	75.1622(12)	90
V (Å^3)	1101.45(6)	2167.7(2)
Z	2	4
ρ_{calc} (g·cm^3)	1.533	1.620
T (K)	200	200
μ (mm^{-1})	0.281	0.655
$2\Theta_{max}$ (°)	28.304	24.812
Total Reflections	15900	25114
Unique Reflections	5341	3671
R$_1$, wR$_2$ (on F^2)	0.0575, 0.1116	0.0938, 0.1146

Table A1.4 Selected bond lengths[a] from the asymmetric units of **2-2** and **2-3** in Chapter 2.

2-2 (Py$_2$TTA)	Distance (Å)	**2-3** (Th$_2$TTA)	Distance (Å)
S1–S4	2.647(1)	S1–S2	2.5922(6)
N1–N4	3.060(4)	N1–N6	2.992(2)
N3–N6	2.998(4)	N3–N8	3.025(2)
C1–C11	3.233(5)	C1–C13	3.233(5)
C2–C12	3.190(4)	C2–C14	3.190(4)
C7–C17	3.480(5)	C4–C15	3.408(2)
S1–N4	3.304(1)	C8–C21	3.346(2)
S1–N6	3.264(2)	S1–N6	3.230(2)
S4–N1	3.252(2)	S1–N8	3.234(1)
S4–N3	3.243(3)	S2–N1	3.234(1)
N4–C1	3.330(5)	S2–N3	3.250(1)
N6–C2	3.190(5)	N6–C1	3.299(2)

Table A1.5 Crystallographic data and selected data collection parameters for **3-1a,b** in Chapter 3.

	3-1a (M = Dy^{III})	**3-1b** (M = Y^{III})
Formula	$C_{14}H_{13}DyN_8O_8S$	$C_{14}H_{13}YN_8O_8S$
Formula Weight	615.88	542.29
Crystal System	Triclinic	Triclinic
Space Group	$P\bar{1}$	$P\bar{1}$
a (Å)	9.3478(3)	9.347(2)
b (Å)	10.5315(4)	10.531(2)
c (Å)	11.3151(4)	11.323(2)
α (°)	69.422(2)	69.422(5)
β (°)	77.825(2)	77.684(5)
γ (°)	77.336(2)	77.232(4)
V (Å^3)	1006.50(6)	1006.3(4)
Z	2	2
ρ_{calc} (g·cm^3)	2.032	1.790
T (K)	200(2)	200(2)
μ (mm^{-1})	3.878	3.064
$2\Theta_{max}$ (°)	28.325	28.479
Total Reflections	13612	4842
Unique Reflections	4926	4842
R_1, wR_2 (on F^2)	0.0428, 0.0863	0.1191, 0.1810

Table A1.6 Selected bond lengths[a] from the asymmetric units of **3-1a** and **3-1b** in Chapter 3.

3-1a	Distance (Å)	3-1b	Distance (Å)
Dy1–O1	2.293(3)	Y1–O1	2.276(5)
Dy1–O8	2.411(3)	Y1–O8	2.399(5)
Dy1–O6	2.444(3)	Y1–O6	2.434(4)
Dy1–O5	2.452(3)	Y1–O5	2.435(4)
Dy1–N1	2.452(3)	Y1–N1	2.442(5)
Dy1–O3	2.463(3)	Y1–O3	2.456(5)
Dy1–N5	2.478(3)	Y1–N4	2.471(5)
Dy1–N4	2.479(4)	Y1–N5	2.473(5)
Dy1–O2	2.497(3)	Y1–O2	2.490(5)
Dy1–N7	2.851(3)	Y1–N7	2.834(5)
Dy1–N6	2.893(4)	Y1–N6	2.876(7)

Table A1.7 Crystallographic data and selected data collection parameters for **4-17** and **4-19** in Chapter 4.

	4-17	**4-19**
Formula	$C_{27}H_{25}N_8S_3Cl$	$C_{17}H_9N_5O_3F_4S_5$
Formula Weight	593.18	567.59
Crystal System	Orthorhombic	Monoclinic
Space Group	$Pbcn$	$P2_1/n$
a (Å)	24.3262(10)	9.0212(17)
b (Å)	8.7298(3)	16.980(3)
c (Å)	27.2563(11)	14.597(3)
α (°)	90	90
β (°)	90	101.067(8)
γ (°)	90	90
V (Å^3)	5788.2(4)	2194.4(8)
Z	8	4
ρ_{calc} (g·cm^3)	1.361	1.718
T (K)	200(2)	201(2)
μ (mm^{-1})	0.381	0.595
$2\Theta_{max}$ (°)	23.545	25.037
Total Reflections	11038	7198
Unique Reflections	11038	3808
R_1, wR_2 (on F^2)	0.0639, 0.1309	0.0471, 0.1209

Table A1.8 Crystallographic data for all phases of **4-7** in Chapter 4.

	α-**4-7**	β-**4-7**	γ-**4-7**	**4-7**·DCE
Formula	$C_{16}H_{10}N_5S_4$	$C_{16}H_{10}N_5S_4$	$C_{16}H_{10}N_5S_4$	$C_{17}H_{12}N_5S_4Cl$
Formula Weight	400.53	400.53	400.53	450.01
Crystal System	Orthorhombic	Triclinic	Monoclinic	Monoclinic
Space Group	$Pbcn$	$P\bar{1}$	$P2_1/c$	$P2_1/n$
a (Å)	7.7256(5)	7.7219(6)	5.4103(5)	8.1755(4)
b (Å)	15.7378(10)	9.3843(8)	24.452(2)	11.2362(9)
c (Å)	26.8749(15)	12.8073(9)	12.9959(13)	21.0981(16)
α (°)	90	105.346(7)	90	90
β (°)	90	90.349(6)	100.024(3)	100.439(4)
γ (°)	90	109.584(7)	90	90
V (Å^3)	3267.6(3)	838.61(12)	1693.0(3)	1906.0(2)
Z	8	2	4	4
ρ_{calc} (g·cm^3)	0.592	0.576	0.571	0.652
T (K)	213(2)	293(2)	200(2)	298(2)
μ (mm^{-1})	0.592	0.576	0.571	0.652
$2\Theta_{max}$ (°)	24.169	22.218	28.449	26.368
Total Reflections	14864	8929	20399	9426
Unique Reflections	2602	2050	4250	3862
R_1, wR_2 (on F^2)	0.0701, 0.1534	0.0433, 0.0993	0.0476, 0.0934	0.0679, 0.1723

Table A1.9 Crystallographic data and selected data collection parameters for β-**4-8** and β-**4-9** in Chapter 4.

	β-**4-8**	β-**4-9**
Formula	$C_{16}H_9N_5S_4F$	$C_{16}H_9N_5S_4Cl$
Formula Weight	418.52	434.97
Crystal System	Monoclinic	Monoclinic
Space Group	$P2_1/c$	$P2_1/n$
a (Å)	21.925(5)	3.9005(6)
b (Å)	5.0789(10)	14.506(2)
c (Å)	15.037(3)	29.824(4)
α (°)	90	90
β (°)	96.840(7)	91.329(5)
γ (°)	90	90
V (Å^3)	1662.6(6)	1686.9(4)
Z	4	4
ρ_{calc} (g·cm^3)	2.032	1.790
T (K)	213(2)	200(2)
μ (mm^{-1})	0.594	0.734
$2\Theta_{max}$ (°)	26.458	24.711
Total Reflections	22184	20524
Unique Reflections	3434	2886
R_1, wR_2 (on F^2)	0.0428, 0.0758	0.0499, 0.0867

Table A1.10 Crystallographic data and selected data collection parameters for α- and β-**5-2** in Chapter 5.

	α-**5-2**[a]	α-**5-2**	β-**5-2**	β-**5-2**
Formula	$C_{12}H_8N_3S_4$	$C_{12}H_8N_3S_4$	$C_{12}H_8N_3S_4$	$C_{12}H_8N_3S_4$
Formula Weight	322.45	322.45	322.45	322.45
Crystal System	Trigonal	Trigonal	Orthorhombic	Orthorhombic
Space Group	$P3_121$	$P3_121$	$Pca2_1$	$Pca2_1$
a (Å)	16.182(3)	16.22280(10)	21.084(4)	20.9556(6)
b (Å)	16.182(3)	16.22280(10)	4.0042(7)	3.8859(1)
c (Å)	4.2947(12)	4.17550(10)	30.170(5)	30.0859(9)
α (°)	90	90	90	90
β (°)	90	90	90	90
γ (°)	120	120	90	90
V (Å^3)	974.0(4)	951.68(3)	2547.1(8)	2449.94(12)
Z	3	3	8	8
ρ_{calc} (g·cm^3)	1.649	1.688	1.682	1.748
T (K)	298(2)	100(2)	293(2)	100(2)
μ (mm^{-1})	0.718	3.660	0.675	7.015
$2\Theta_{max}$ (°)	27.50	25.491	25.568	66.577
Total Reflections	8921	12312	5095	13465
Unique Reflections	1500	1304	5095	3847
R_1, wR_2 (on F^2)	0.0363, 0.0725	0.0166, 0.0461	0.0651, 0.1632	0.0344, 0.0813

[a] From ref 29 in Chapter 5.

A.2 High Temperature Series Expansion Function for α-PhBPMe

Based on the expanded trigonal cell diagram shown in **Figure 5.9**, a fit function to estimate for the magnetic susceptibility χ of the trigonal phase of PhBPMe in terms of the unique pairwise exchange energies J_1, J_2, and J_3 $(= J_\pi)$ was developed up to eighth order in $1/T$ ($n = 8$ below) according to method described by Schmidt, Lohmann and Richter (ref. 50 in Chapter 5). The general expression is given by:

$$\chi = C \sum_n \frac{c_n}{(k_b T)^n}$$

where C is the Curie constant, and the coefficients c_n are given in terms of J_1, J_2, and J_3 $(= J_\pi)$.

Note: In the expressions for c_n provided below, J-values are defined with respect to the Hamiltonian $\widehat{\mathcal{H}}_{ex} = J_{ij}\langle \widehat{\mathbf{S}}_i \cdot \widehat{\mathbf{S}}_j \rangle$. The values extracted from a fit to the data using these expressions were adjusted afterwards to conform with the convention $\widehat{\mathcal{H}}_{ex} = -2J_{ij}\langle \widehat{\mathbf{S}}_i \cdot \widehat{\mathbf{S}}_j \rangle$ used in the main text.

```
c1 = 0.2500000000

c2 =-2.500000000000e-01*J1 - 2.500000000000e-01*J2 - 1.250000000000e-01*J3

c3 = 1.250000000000e-01*J1**2 + 5.000000000000e-01*J1*J2 +
1.250000000000e-01*J2**2 + 2.500000000000e-01*J1*J3 +
2.500000000000e-01*J2*J3

c4 = -4.166666666667e-02*J1**3 - 4.348958333333e-01*J1**2*J2 -
4.739583333333e-01*J1*J2**2 - 4.166666666667e-02*J2**3 -
2.500000000000e-01*J1**2*J3 - 6.197916666667e-01*J1*J2*J3 -
2.500000000000e-01*J2**2*J3 - 6.250000000000e-02*J1*J3**2 -
6.250000000000e-02*J2*J3**2 + 1.041666666667e-02*J3**3

c5 = 2.994791666667e-02*J1**4 + 2.269965277778e-01+J1**3*J2 +
7.154947916667e-01*J1**2*J2**2 + 2.812500000000e-01*J1*J2**3 +
2.994791666667e-02*J2**4 + 1.406250000000e-01*J1**3*J3 +
8.190104166667e-01*J1**2*J2*J3 + 8.190104166667e-01*J1*J2**2*J3 +
1.536458333333e-01*J2**3*J3 + 1.562500000000e-01*J1**2*J3**2 +
2.773437500000e-01*J1*J2*J3**2 + 1.562500000000e-01*J2**2*J3**2 -
2.083333333333e-02*J1*J3**3 - 2.083333333333e-02*J2*J3**3 +
3.255208333333e-03*J3**4

c6 = -2.356770833333e-02*J1**5 - 1.353624131944e-01*J1**4*J2 -
6.078559027778e-01*J1**3*J2**2 - 7.245008680556e-01*J1**2*J2**3 -
1.622178819444e-01*J1*J2**4 - 2.356770833333e-02*J2**5 -
6.835937500000e-02*J1**4*J3 - 7.163628472222e-01*J1**3*J2*J3 -
1.419487847222e+00*J1**2*J2**2*J3 - 7.693142361111e-01*J1*J2**3*J3 -
8.018663194444e-02*J2**4*J3 - 1.777343750000e-01*J1**3*J3**2 -
5.970052083333e-01*J1**2*J2*J3**2 - 5.891927083333e-01*J1*J2**2*J3**2 -
1.796875000000e-01*J2**3*J3**2 - 1.041666666667e-02*J1**2*J3**3 -
8.680555555556e-03*J1*J2*J3**3 - 1.041666666667e-02*J2**2*J3**3 +
3.906250000000e-03*J1*J3**4 + 3.906250000000e-03*J2*J3**4 -
1.367187500000e-03*J3**5
```

```
c7 = 4.871961805556e-03*J1**6 + 1.184678819444e-01*J1**5*J2 +
3.957302517361e-01*J1**4*J2**2 + 9.856987847222e-01*J1**3*J2**3 +
5.382541232639e-01*J1**2*J2**4 + 1.241699218750e-01*J1*J2**5 +
6.092664930556e-03*J2**6 + 6.080729166667e-02*J1**5*J3 +
4.917263454861e-01*J1**4*J2*J3 + 1.740272352431e+00*J1**3*J2**2*J3 +
1.789621310764e+00*J1**2*J2**3*J3 + 5.953667534722e-01*J1*J2**4*J3 +
6.093750000000e-02*J2**5*J3 + 1.187825520833e-01*J1**4*J3**2 +
8.188368055556e-01*J1**3*J2*J3**2 + 1.304215494792e+00*J1**2*J2**2*J3**2 +
8.129665798611e-01*J1*J2**3*J3**2 + 1.282280815972e-01*J2**4*J3**2 +
7.816840277778e-02*J1**3*J3**3 + 1.649305555556e-01*J1**2*J2*J3**3 +
1.687391493056e-01*J1*J2**2*J3**3 + 7.351345486111e-02*J2**3*J3**3 -
2.073567708333e-02*J1**2*J3**4 - 2.479383680556e-02*J1*J2*J3**4 -
2.073567708333e-02*J2**2*J3**4 + 5.989583333333e-03*J1*J3**5 +
5.989583333333e-03*J2*J3**5 -1.082356770833e-03*J3**6

c8 = 6.828187003968e-04*J1**7 - 5.989990234375e-02*J1**6*J2 -
3.101241500289e-01*J1**5*J2**2 - 8.564520941840e-01*J1**4*J2**3 -
1.073144079138e+00*J1**3*J2**4 - 3.742840802228e-01*J1**2*J2**5 -
7.537887008102e-02*J1*J2**6 - 2.354213169643e-04*J2**7 -
4.366590711806e-02*J1**6*J3 - 3.521421079282e-01*J1**5*J2*J3 -
1.631472891348e+00*J1**4*J2**2*J3 - 2.810948350694e+00*J1**3*J2**3*J3 -
1.842822265625e+00*J1**2*J2**4*J3 - 4.239859121817e-01*J1*J2**5*J3 -
4.374547887731e-02*J2**6*J3 - 7.346462673611e-02*J1**5*J3**2 -
8.087391040943e-01*J1**4*J2*J3**2 - 2.089635326244e+00*J1**3*J2**2*J3**2 -
2.097923448351e+00*J1**2*J2**3*J3**2 - 8.452121310764e-01*J1*J2**4*J3**2 -
8.492725513600e-02*J2**5*J3**2 - 1.097819010417e-01*J1**4*J3**3 -
4.194905598958e-01*J1**3*J2*J3**3 - 6.256347656250e-01*J1**2*J2**2*J3**3 -
4.042932581019e-01*J1*J2**3*J3**3 - 1.045875831887e-01*J2**4*J3**3 +
1.278754340278e-02*J1**3*J3**4 + 2.748390480324e-02*J1**2*J2*J3**4 +
2.600685402199e-02*J1*J2**2*J3**4 + 1.231418185764e-02*J2**3*J3**4 -
2.522786458333e-03*J1**2*J3**5 - 4.423466435185e-03*J1*J2*J3**5 -
2.522786458333e-03*J2**2*J3**5 + 3.634982638889e-04*J1*J3**6 +
3.634982638889e-04*J2*J3**6 + 6.200396825397e-05*J3**7
```

$$\chi_{\text{HTSE}} = c_1/T + c_2/T + c_3/T + c_4/T + c_5/T + c_6/T + c_7/T + c_8/T$$

$$\chi = \text{tip} + 4*C*\chi_{\text{HTSE}}$$

A.3 Permissions Granted for the Reproduction of Published Content in Thesis

A3.1.1 Chapter 2

Adapted with permission from *Cryst. Growth Des.*, 2015, **15**, 2524–2532. Copyright 2020 American Chemical Society.

Adapted with permission from *Inorg. Chem.*, 2019, **58**, 419–427. Copyright 2020 American Chemical Society.